Solutions and Tests

for

Exploring Creation

with

Chemistry
2nd Edition

by Dr. Jay L. Wile

Solutions and Tests for Exploring Creation With Chemistry, 2nd Edition

Published by
Apologia Educational Ministries, Inc.
1106 Meridian Plaza, Suite 220
Anderson, IN 46016
www.apologia.com

Manufactured in the United States of America
Eighth Printing, June 2011

ISBN: 978-1-932012-27-9

Printed by Courier, Inc., Westford, MA

Cover photos © 1999 Photodisc, Inc. *Cover design by Kim Williams*

Any illustration not credited in the text was done by Megan Whitaker

Exploring Creation With Chemistry, 2nd Edition

Solutions and Tests

TABLE OF CONTENTS

Sample Calculations for Selected Experiments:

Solutions to the Extra Practice Problems:

Module Tests:

Solutions to the Module Tests:

Quarterly Tests:

Solutions to the Quarterly Tests:

TEACHER'S NOTES

Exploring Creation With Chemistry, 2nd Edition

Thank you for choosing *Exploring Creation With Chemistry.* I designed this course to meet the needs of the homeschooling parent. I am very sensitive to the fact that most homeschooling parents do not know chemistry very well, if at all. As a result, they consider it nearly impossible to teach to their children. This course has several features that make it ideal for such a parent:

1. The course is written in a conversational style. Unlike many authors, I do not get wrapped up in the desire to write formally. As a result, the text is easy to read, and the student feels more like he or she is *learning*, not just reading.

2. The course is completely self-contained. Each module in the student text includes the text of the lesson, experiments to perform, problems to work, and questions to answer. This book contains the worked-out solutions to the questions and problems in the student text, tests, worked-out solutions to the tests, and some extra material (example calculations for several experiments as well as cumulative tests and their worked-out solutions).

3. The chemicals for the experiments are readily available at either the grocery or the hardware store. In addition, nearly all of the experiments can be performed with household equipment such as glasses, measuring cups, spoons, etc. If you wish to perform every experiment contained in the course, however, I suggest that you purchase a set of equipment that we make available for this course. A list of the equipment is presented in the "Student Notes" section of the student text. In addition, there are some extra experiments that are referenced in the course. If your student enjoys experiments, you can purchase a kit that will allow you to perform those experiments as well. The details are discussed in the "Student Notes" section of the student text.

4. Most importantly, this course is Christ-centered. In every way possible, I try to make the science of chemistry glorify God. One of the most important things that you and your student should get out of this course is a deeper appreciation for the wonder of God's creation!

Pedagogy of the Text

There are three types of exercises that the student is expected to complete: "On Your Own" problems, review questions, and practice problems.

- The "On Your Own" problems should be solved as the student reads the text. The act of working out these problems will cement in the student's mind the concepts he or she is trying to learn. The solutions to these problems are included as a part of the student text. The student should feel free to use those solutions to check his or her work.

- The review questions are conceptual in nature and should be answered after the student completes the module. They will help the student recall the important concepts from the reading. As your student's teacher, you can decide whether or not your student can look at the solutions to these questions. They are located in this book.

- The practice problems should also be solved after the module has been completed, allowing the student to review the important quantitative skills from the module. As your student's teacher, you can decide whether or not your student can look at the solutions to these problems. They are located in this book.

In addition to the problems, there is also a test for each module. Those tests are in this book, but a packet of those tests is also included with this book. You can tear the tests out of the packet and give them to your student so that you need not give him or her this book. You can also purchase additional packets for additional students. You also have our permission to copy the tests out of this book if you would prefer to do that instead of purchasing additional tests for additional students. **I strongly recommend that you administer each test once the student has completed the module and all of the associated exercises. The student should be allowed to have only a calculator, pencil, paper, and a copy of the periodic chart (provided in Module #3 and on the inside cover of the student text) while taking the test.**

There are also cumulative tests in this book. You can decide whether or not to give these tests to your student. Cumulative tests are probably a good idea if your student is planning to go to college, as he or she will need to get used to taking such tests. There are four cumulative tests along with their solutions. Each cumulative test covers four modules. You have three options as to how you can administer them: You can give each test individually so that the student has four quarterly tests. You can combine the first two quarterly tests and the second two quarterly tests to make two semester tests. You can also combine all four tests to make one end of the year test. If you are giving these tests for the purpose of college preparation, I recommend that you give them as two semester tests, because that is what the student will face in college. The cumulative tests are not in the packet of tests. However, you have our permission to copy them out of this book so that you can give them to your student.

Any information that the student must memorize is centered in the text and put in boldface type. Any boldface words (centered or not) are terms with which the student must be familiar. In addition, all definitions presented in the text need to be memorized. Finally, if an equation must be used to answer any "On Your Own" problem, review question, or practice problem, it must be memorized for the test. In general, these student exercises are meant as a study guide for the tests. Skills and knowledge necessary to complete them will be required for the test.

You will notice that every solution contains an underlined section. That is the answer. The rest is simply an explanation of how to get the answer. For questions that require a sentence or paragraph as an answer, the student need not have *exactly* what is in the solution. The basic message of his or her answer, however, has to be the same as the basic message given in the solution. Please note that in the mathematical problems that your student must do, his or her answer need not be *identical* to my answer. It must have the same number of significant figures (you will learn what that means in Module #1) as my answer, but it can vary by one or two digits in the last decimal place. This is to be expected due to rounding errors.

Experiments

The experiments in this course are designed to be done as the student is reading the text. I recommend that your student keep a notebook of these experiments. The details of how to perform the experiments and how to keep a laboratory notebook are discussed in the "Student Notes" section of the student text. If you go to the course website that is discussed in the "Student Notes" section of the student text, you will also find examples of how the student should record his or her experiments in the laboratory notebook.

Grading

Grading your student is an important part of this course. I recommend that you *correct* the review questions and practice problems, but I do not recommend that you include the student's score in his or her grade. Instead, I recommend that the student's grade be composed solely of test grades and laboratory notebook grades. Here is what I suggest you do:

1. Give the student a grade for each lab that is done. This grade should not reflect the accuracy of the student's results. Rather, it should reflect how well the student followed directions, how well the student did the calculations and kept track of significant figures, and how well he or she wrote up the lab in his or her lab notebook. If a lab requires calculations, a sample of those calculations are included in this book. That sample will help you determine whether or not your student did the calculations properly and kept track of significant figures.

2. Give the student a grade for each test. In the test solutions, you will see a point value assigned to each problem. If your student answered the problem correctly, he or she should receive the number of points listed. If your student got a portion of the problem correct, he or she should receive a portion of those points. If your student's answer is correct except for the number of significant figures, you should take off ¼ of a point. Your student's percentage grade, then, can be calculated as follows:

$$\text{Student's Grade} = \frac{\#\text{ of points received}}{\#\text{ of points possible}} \times 100$$

   The number of possible points for each test is listed at the bottom of the solutions.

3. The student's overall grade in the course should be weighted as follows: 35% lab grade and 65% test grade. If you use the cumulative tests, make them worth twice as much as each module test. If you really feel that you must include the review questions and practice problems in the student's total grade, make the labs worth 35%, the tests worth 55%, and the review questions and practice problems worth 10%. A straight 90/80/70/60 scale should be used to calculate the student's letter grade. This is typical for most schools. If you have your own grading system, please feel free to use it. This grading system is only a suggestion.

Finally, I must tell you that I pride myself on the fact that this course is user-friendly and reasonably understandable. At the same time, however, *it is not EASY*. This is a tough course. I have designed it so that any student who gets a "C" or better on the tests will be very well prepared for college.

Question/Answer Service

For all those who use this curriculum, we offer a question/answer service. If there is anything in the modules that you do not understand - from an esoteric concept to a solution for one of the problems - just contact us via any of the methods listed on the **NEED HELP?** page of the student text. You can also contact us regarding any grading issues that you might have. This is our way of helping you and your student to get the maximum benefit from our curriculum.

ANSWERS TO THE REVIEW QUESTIONS FOR MODULE #1

1. Pretty much everything except light contains matter. Since (d) is the only thing on the list that is just light, the lightning bolt (d) has no matter.

2. Length is measured in meters, mass is measured in grams, time is measured in seconds, and volume is measured in liters.

3. The prefix "centi" means 0.01, or one hundredth.

4. Since a kL is equal to 1,000 L and a mL is only equal to 0.001 L, the glass holding 0.5 kL has more liquid in it.

5. The ruler is marked off in 0.1 cm. You can estimate to the next decimal place. Since the end of the ribbon falls between 2.6 cm and 2.7 cm, the best answer is 2.65 cm. Answers from 2.63 to 2.67 should be counted correct as well.

6. Assuming both students reported the proper number of significant figures, the first student was more precise (because that measurement has more places to the right of the decimal), but the second student was more accurate (because that measurement is closer to the correct value).

7. a. The first two zeros are not significant, but the other two are. Thus, there are 6 significant figures.
   b. All of the zeros are significant. Thus, there are 5 significant figures.
   c. There are 2 significant figures.
   d. The zero is significant. Thus, there are 3 significant figures.

8. The student's value for density has far too many significant figures, and it has no units attached. Either of these things would make the student's answer wrong.

9. Ice floats on water because its density is lower than that of water.

10. The golden statue would be heavier because the density of gold is greater than that of lead.

ANSWERS TO THE REVIEW QUESTIONS FOR MODULE #2

1. For the first 1.5 minutes, the man did no work. Motion is required in order to do work. Since the car did not move during that time, no work was done. For the last 0.5 minutes, however, the man did do work.

2. a. A lump of coal has potential energy in it because all matter has stored energy.
   b. A flash of lightning is light, so it has no matter. Thus, it cannot have potential energy. However, the lightning flash is in motion, so it has kinetic energy.
   c. A candle flame also has no matter, but it does have heat, so it has kinetic energy.
   d. A tornado is a swirling mass of air. Since air is matter, it has potential energy, but because the air is moving, it also has kinetic energy. Therefore, it has both.

3. If a substance's temperature decreases, we can conclude that it is losing energy.

4. Since the object cannot interact with anything else, it will never be able to give away any of its energy or absorb energy from anything. Thus, its temperature can never change.

5. Science is fallible. Laws which scientists thought were true for hundreds of years have later been demonstrated to be false. Since science is fallible, it is a poor source for ultimate truth. It is, however, an excellent way to understand and appreciate God's creation.

6. Since it takes 4.184 Joules to make 1 calorie, the Joule is a smaller energy unit. Thus, the object which absorbs 100.0 cal is absorbing more energy and will therefore be the hottest.

7. Your body has an internal temperature of about 37 °C. Ice water has a temperature of 0.0 °C. Your body must therefore warm up the water. That takes energy, which your body gets from the food that you eat. Thus, drinking ice-cold water is a way of burning off excess Calories.

8. In Experiment 2.1, we learned that the heating curve for water has two flat regions. The first occurs when ice melts, and the second occurs when water boils. Based on this fact, the unknown liquid must melt at 5 °C and boil at 55 °C. Your answer can be slightly different than this one, because you are reading it from a graph.

9. Since the iron has a larger specific heat, it is harder to heat up. Thus, the gold will come out hottest. Since gold's specific heat is three times smaller than iron, we can actually conclude that it will come out three times hotter. Thus, if the gold increased in temperature by 900 °C, the iron would only increase in temperature by 300 °C.

10. A calorimeter should be made out of insulating material because energy should never be allowed to escape from it. If energy does escape from a calorimeter, the measurement you are using the calorimeter for will be in error.

ANSWERS TO THE REVIEW QUESTIONS FOR MODULE #3

1. The continuous theory of matter states that matter comes in long, continuous sheets. The discontinuous theory, however, assumes that matter come in little packets and the only reason matter looks continuous is that we cannot magnify it enough to see the little packets of matter.

2. The Law of Mass Conservation and the Law of Definite Proportions were instrumental in the development of Dalton's atomic theory. Dalton predicted the Law of Multiple Proportions.

3. Matter cannot be created or destroyed. Thus, throughout any change that occurs in matter, the total mass before the change must equal the total mass after the change.

4. Dalton's atomic theory assumed four things:

   a. All elements are composed of small, indivisible particles called "atoms."

   b. All atoms of the same element have exactly the same properties.

   c. Atoms that make up different elements have different properties.

   d. Compounds are formed when atoms are joined together. Since atoms are indivisible, they can only join together in simple, whole number ratios.

5. An atom is the smallest unit of matter. A molecule is also a unit of matter, but is formed when atoms join together. Thus, a molecule is made up of more than one atom.

6. Metals are malleable, have luster, and conduct electricity. Nonmetals are brittle, lack luster, and do not conduct electricity.

7. There is a heavy, jagged line that runs down the right side of the chart. If an atom lies to the left of that line (excluding hydrogen), it is a metal. If it lies to the right of that line, or if it is hydrogen, it is a nonmetal.

8. A compound is ionic if, when dissolved in water, it conducts electricity. If it does not conduct electricity, the compound is covalent. Thus, you can perform experiments like Experiment 3.2 to determine whether a compound conducts electricity when dissolved in pure water. That will determine whether it is ionic or covalent.

9. If a compound has a metal in it, it must be ionic. If it has no metals, it is covalent.

10. The way atoms can join together is different between ionic and covalent compounds. In ionic compounds there is only one possible combination of atoms. In covalent compounds, many combinations are possible. Therefore, we need 2 naming systems.

ANSWERS TO THE REVIEW QUESTIONS FOR MODULE #4

1. Anything that can be separated into its components must be a mixture. Compounds must first be *decomposed* before they can be separated into their component elements.

2. Anything that has a single chemical name is made up of a single atom or molecule. As a result, it must be a pure substance.

3. Nitrogen makes up 78% of the air we breathe.

4. Once again, the answer is Nitrogen. Since we don't use up the nitrogen we breathe in, it also makes up the vast majority of what we exhale. You may have learned in biology that we breathe in oxygen and breathe out carbon dioxide. This is true, but oxygen is only 21% of what we breathe in and carbon dioxide is only 4% of what we breathe out!

5. Since molecules move faster and are farther apart in the gas phase compared to the solid phase, a transition from solid to gas must occur when a substance is heated.

6. The only phase in which molecules move slower and are closer together than in the liquid phase is the solid phase.

7. Water expands when it freezes; thus, its molecules get farther apart. All other natural substances contract when they freeze because their molecules get closer together.

8. A chemical change alters the identity of the substances involved in a change. A physical change does not.

9. The homonuclear diatomics are N_2, O_2, Cl_2, F_2, Br_2, I_2, At_2, and H_2.

10. A chemical equation is balanced if each type of atom in the equation is present in the same number on both sides of the arrow.

ANSWERS TO THE REVIEW QUESTIONS FOR MODULE #5

1. Elements cannot undergo decomposition reactions, because elements have nothing to decompose into.

2. a. This equation represents a complete combustion reaction, because O_2 is added while CO_2 and H_2O are produced.
   b. This equation represents a formation reaction, because 2 elements come together to produce 1 compound.
   c. None of these
   d. None of these
   e. This equation represents a decomposition reaction, because a single compound is breaking down into it elements.

3. The only equation that does not represent a formation reaction is (a). The equation for (b) is a formation equation, because it starts with two reactants and makes only one in the end. Remember, reactions do not have to start with elements in order to be formation equations, we just assume they do when we write them ourselves.

4. The only equation that does not represent complete combustion is (b). The equation in (b) is a decomposition equation. Even though it produces CO_2 and H_2O, it does not have O_2 as a reactant, so it is not combustion.

5. Complete combustion produces carbon dioxide and water. Incomplete combustion produces either carbon monoxide and water or carbon and water.

6. A catalytic converter converts carbon monoxide produced in an automobile to carbon dioxide, a non-poisonous gas.

7. According to the chart, 100 H's would have a mass of 100 x 1.01 = 101 amu; 4 S's would have a mass of 4 x 32.1 = 128.4 amu, and 1 La has a mass of 138.9 amu. Thus 1 La atom has the most mass.

8. All of the samples contain one mole.

9. Avogadro's Number is 6.02×10^{23}, and it represents the number of things that are contained in one mole.

10. In stoichiometry, we convert the quantity of one substance in a chemical reaction into the quantity of another substance. Until the equation is balanced, however, it does not properly represent a chemical reaction, so it is useless until it is balanced.

ANSWERS TO THE REVIEW QUESTIONS FOR MODULE #6

1. Stoichiometry is the method used to relate the quantities of substances in chemical equations. It is useful because with the quantity of just one substance in the equation, you can learn something about the quantities of all other substances in that equation.

2. A limiting reactant is the reactant in a chemical equation that runs out first. It is important because its quantity determines the amount of products produced.

3. The limiting reactant is $K_2Cr_2O_7$ because 14 moles of HCl react with 1 mole of $K_2Cr_2O_7$. Since 15 moles of HCl are added, the 1 mole of $K_2Cr_2O_7$ will run out and there will still be 1 extra mole of HCl.

4. Gay Lussac's Law says that the stoichiometric coefficients in a chemical equation relate the volume of gases in that equation as well as the number of moles. It can be used to relate the quantities of gaseous substances in a chemical equation if those quantities are given in volume units.

5. The chemist can only use Gay-Lussac's Law to relate the volumes of HCl and CO_2, since those are the only gases in the equation.

6. Since stoichiometry can only be done in moles (or volume in the case of gases), the first thing you must do is convert to moles.

7. Molecular formulas provide the exact number of each type of atom in the molecule. Empirical formulas provide only a simple, whole-number ratio of atoms in the molecule.

8. The formulas given in (b) and (c) are empirical formulas, because the subscripts have no common factor. The formula given in (a) is not an empirical formula because the subscripts have a common factor of 2.

9. The common factor between the subscripts is 7; thus we must divide each subscript by 7 to get C_2H_3O.

10. The molecular formula is also the empirical formula, since the subscripts have no common factor. Thus, the empirical formula is also H_3PO_4.

ANSWERS TO THE REVIEW QUESTIONS FOR MODULE #7

1. Rutherford experimentally determined that the plum pudding model of the atom was incorrect and proposed his own model of the atom, which we call the planetary model. It became the foundation upon which Bohr's model was built.

2. It is called a Crookes Tube, and he used it to discover cathode rays, which were later determined to be electrons.

3. Like charges repel each other. Thus, the particles must have the same type of charge.

4. If a substance has an imbalance of charges, it takes on the charge that is more plentiful. Thus, this substance will be positively charged.

5. Protons and neutrons are tightly packed together in the nucleus of the atom.

6. Isotopes behave identically in terms of chemistry. It is therefore nearly impossible to separate them.

7. The plum pudding model of the atom had the positive and negative charges equally disbursed throughout the entire atom. The planetary model, on the other hand, concentrated the positive charges at the center of the atom and had the negative charges whirling around on the outside.

8. Remember ROY G. BIV. This is the order of visible light wavelengths from the largest to the smallest. Thus, the orange light bulb has larger wavelengths. When wavelength is large, however, frequency is small; thus, the violet light has the highest frequency. The higher the frequency, the higher the energy, so the violet light also has the highest energy.

9. The wavelengths emitted by the lights are the same, but the brighter bulb emits waves of larger amplitude.

10. When atoms absorb energy, their electrons jump to higher energy orbitals. When they emit light, the electrons are dropping down into lower energy orbitals.

11. The ground state of any substance is its lowest possible energy state. This is important in chemistry because all matter strives to reach its ground state.

12. The neutron is the heaviest, the proton is next, and the electron is the lightest. The proton and neutron differ only slightly in mass, but the proton is 2,000 times heavier than the electron.

ANSWERS TO THE REVIEW QUESTIONS FOR MODULE #8

1. Valence electrons are those electrons farthest away from the nucleus of an atom. They are important in chemistry because they are responsible for determining the chemical behavior of an atom.

2. The valence electrons in this configuration are the ones in the fifth energy level, since this is the farthest from the nucleus. There are a total of 5 (3 + 2) electrons in the fifth energy level, so there are 5 valence electrons.

3. Since life is based on carbon, it could, in theory, be based on any atom in the same column, since they all have similar chemistry. Thus, life could be based on Si, Ge, Sn, or Pb as well.

4. The noble gases are the elements in group 8A. They are important because they have ideal electron configurations.

5. Metals tend to give up electrons to attain the ideal electron configuration, while nonmetals tend to gain electrons for the same purpose.

6. An atom has no electrical charge. Ions have electrical charge because they have an imbalance of protons and electrons.

7. Ionization potential is the energy required to remove an electron from an atom.

8. Ionization potential, electronegativity, and atomic radius are all periodic properties.

9. Ionic compounds form when atoms give and take electrons. Thus, they are composed of ions. Covalent compounds result when atoms share electrons. Thus, no ions are formed.

10. Nitrogen, oxygen, and ozone protect the earth from the harmful rays of the sun.

ANSWERS TO THE REVIEW QUESTIONS FOR MODULE #9

1. All of the ions we have learned about up to this point have consisted of only one atom. The ions we learned about in this module are made up of more than one atom.

2. You should have all of these memorized:

   a. SO_4^{2-}

   b. NO_3^-

   c. $C_2H_3O_2^-$

   d. This is a question that tests your memory. Sulfide is the name of the ion formed from a single sulfur ion. As you learned in Module #8, when a single atom forms a negative ion, the ion's name is the name of the atom with an "ide" ending. Since sulfur is in group 6A on the chart, it takes on a -2 charge in ionic compounds. Thus, the sulfide ion is S^{2-}.

   e. OH^-

   f. PO_4^{3-}

3. Once again, these should be memorized:

   a. nitrite

   b. This is also a question to test your memory. This is the oxide ion. You learned in Module #8 that when a single atom forms a negative ion, the ion's name is the name of the atom with an "ide" ending.

   c. chlorate

   d. carbonate

   e. sulfite

   f. chromate

4. VSEPR stands for Valence Shell Electron Pair Repulsion.

5. VSEPR theory states that a molecule will take on whatever shape gets the central atom's valence electron pairs as far away from each other as possible.

6. Bond angles in a pyramidal molecule are smaller than those in a tetrahedral molecule because the non-bonding electron pair in the pyramidal molecule pushes the bonding electron pairs a little harder than they push on each other. As a result, the bonds are all pushed towards each other a little more than in the tetrahedral molecule, which has no non-bonding electron pairs.

7. Polar covalent bonds are bonds in which the electrons are not shared equally. Purely covalent bonds, on the other hand, contain electrons that are shared perfectly equally.

8. Their electronegativities must be the same. After all, if the bond is purely covalent, the electron pair is shared equally. This can only happen if the atoms tug on the electrons with the same strength. Since that strength is determined by electronegativity, the electronegativities must be the same. Do not think that different atoms must have different electronegativities. Electronegativity increases as you go from left to right on the chart, and it decreases as you go down the chart. Thus, if one atom is both to the right and *below* another atom, those two atoms could have the same electronegativity. At and H, for example, have identical electronegativities.

9. Oil and water do not mix because oil is a purely covalent compound and water is a polar covalent compound. These two types of compounds cannot mix because one has electrical charges and the other doesn't.

10. Soap is made up of long molecules that are ionic on one end and purely covalent on the other. Thus, they are attracted to both the purely covalent molecules that make up the stain and the polar covalent water molecules.

ANSWERS TO THE REVIEW QUESTIONS FOR MODULE #10

1. a. Bases usually taste bitter, so this is probably a base.
   b. Bases usually feel slippery, so it is probably a base.
   c. Acids turn blue litmus red, so this is an acid.

2. An acid is an H^+ donor. In this equation, the CH_4O turned into CH_3O^-, indicating that it lost an H^+ ion. Thus, CH_4O is the acid.

3. We defined a base as an H^+ acceptor. But, as discussed in the module, an H^+ ion is simply a proton. Therefore, the definitions are the same.

4. Polyprotic acids have more than one H^+ to donate. Thus, (b) and (d) are polyprotic acids. Although (c) has more than one H in its formula, it is a base, not an acid.

5. Concentration refers to the amount of a substance (grams, moles) divided by volume (m^3, liters, etc.). Only (b) and (d) have units that involve an amount unit divided by a volume unit.

6. In order to make a new HCl solution from an old one, the chemist must perform a dilution. The only thing you can do in a dilution is *lower* the concentration from the original. Thus, the chemist must use the 6.0 M solution because only this one has a higher concentration than the one that the chemist wants to make.

7. You should recognize the carbonate ion in this compound. The metal forms the positive ion. Thus, this compound splits up into two Al^{3+} ions and three CO_3^{2-} ions.

8. An amphiprotic substance can act like both an acid and a base. In these two equations, HCO_3^- is doing this.

9. An indicator can determine whether a substance is an acid or base. It can also help you find the endpoint of a titration.

10. The endpoint of a titration tells you that you have added just enough acid in your titration to eat up all of the base in your unknown solution. Alternatively, it can tell you that you have added just enough base to your titration to eat up all of the acid in your unknown solution, depending on whether your unknown is an acid or a base.

ANSWERS TO THE REVIEW QUESTIONS FOR MODULE #11

1. The solute is what you are dissolving, and the solvent is what you are dissolving the solute into. Therefore, salt is the solute, and water is the solvent.

2. Ionic compounds split up into their ions when they dissolve. Thus, an ionic compound will split up into 2 or more ions in solution. On the other hand, polar covalent compounds dissolve one molecule at a time.

3. A saturated solution contains as much solute as is possible for the given temperature and pressure.

4. Solids dissolve best at high temperature.

5. Under high pressure conditions, gases dissolve best.

6. Liquid solutes are not affected strongly by the conditions under which the solution is made.

7. To increase the solubility of a solid, you must increase the temperature of the solvent.

8. When something dissolves endothermically, it cools the solution. You should expect the beaker to get cold.

9. Molality takes number of moles of solute and divides by the *kilograms* of *solvent.* Molarity takes the number of moles of solute and divides by the *liters* of the *solution*.

10. The only solute characteristic that affects the freezing point depression of a solvent is the number of molecules (or ions) it splits up into when it dissolves. The more molecules (or ions) the solute splits into, the larger the value of "i." The larger the value of "i," the larger the freezing point depression. $Al(NO_3)_3$ splits into 4 ions, so it does the best job of protecting water from freezing.

ANSWERS TO THE REVIEW QUESTIONS FOR MODULE #12

1. Pressure is defined as the force per unit area that a gas exerts on its surroundings. The units used to measure pressure are Pa, kPa, atms, torr, and mmHg. The student need only name three of these units.

2. Boyle's Law states that under conditions of constant temperature, the product of a gas's pressure and volume is always constant.

3. The Kelvin temperature scale derives from Charles's Law. It is the result of the fact that when you extrapolate the volume versus temperature data for all gases, zero volume occurs at -273.15 °C. This means that nothing can ever get colder than -273.15 °C, or 0.00 K.

4. A careful scientist can only extrapolate data when the amount of data is large compared to the extrapolation.

5. The molecules (or atoms) that make up an ideal gas must be small compared to the volume available to the gas. The gas molecules (or atoms) must be so far apart that they do not attract or repel each other. Also, all collisions that occur must be elastic.

6. Gases behave ideally when their pressure is near or lower than 1.00 atm and when their temperature is near or higher than 273K.

7. STP is defined as 1.00 atm and 273 K.

8. They both have exactly the same pressure, because Dalton's Law tells us that the pressure of an ideal gas is independent of its identity.

9. When the temperature of a liquid is lowered, its vapor pressure lowers. Thus, the vapor pressure decreased.

10. Mole fraction represents the fraction of molecules represented by the component of interest. Thus, a mole fraction of 0.78 tells us that for every 100 molecules, 78 will be nitrogen. Thus, for every 1,000 molecules, 780 will be nitrogen.

ANSWERS TO THE REVIEW QUESTIONS FOR MODULE #13

1. Potential energy is stored in the bonds of the molecules in the reaction. Kinetic energy is the heat that is either released or absorbed in the reaction.

2. A positive ΔH indicates an endothermic reaction. Endothermic reactions absorb heat from their surroundings. This leaves less energy for the surroundings. As a result, the beaker will feel cold.

3. Hess's Law is more exact because it also takes into account the phases of each substance in the reaction.

4. A state function is one whose final value is independent of path. Enthalpy and Gibbs Free Energy are examples of state functions.

5. Only elements that are in their elemental form have a ΔH_f^o of zero. NaOH is not an element. Na^+ is not an element either; it is an ion. O is an element, but the elemental form of oxygen is gaseous O_2; thus, the ΔH_f^o for O is not zero. Finally, H_2 (l) is not the elemental form of hydrogen, H_2 (g) is. Thus, only O_2 (g) and Cl_2 (g) have ΔH_f^o's of zero.

6. Endothermic reactions have their reactants at a lower energy than their products. Thus, diagram I is the endothermic reaction. Additionally, the ΔH of the reaction is just the energy of the products minus that of the reactants. Thus ΔH = 70 kJ - 18 kJ = 52 kJ. Since the answer is obtained by reading a graph, the answer can be anywhere from 52 kJ to 57 kJ.

7. If ΔH is zero, the reactant and products have the same potential energy. A large activation energy means a large hump at the intermediate state. The drawing on the right, then, represents a reaction in which the ΔH is zero and the activation energy is large.

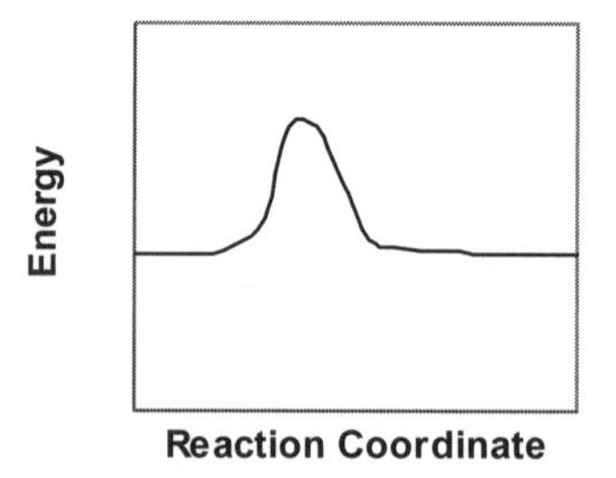

8. Increased temperature means increase molecular (or atomic) motion and thus more disorder. Therefore, the warmer block has a higher entropy.

9. When you study, your brain's entropy does decrease, but at the same time, your body is burning fuel to give you the energy you need to study. This process heats up the surroundings, causing entropy to increase. The entropy increase of the surroundings offsets the entropy decrease of your brain, and thus the change in entropy of the universe is still positive, in agreement with the Second Law.

10. If a reaction is exothermic, the first term in Equation (13.13) is negative. The problem is that the second term is also negative if ΔS is negative. Thus, you have one negative number being subtracted from another negative number. In order to ensure that the result is negative (and thus the reaction is spontaneous), you must decrease the TΔS term. The only way you can do that is to lower the temperature.

ANSWERS TO THE REVIEW QUESTIONS FOR MODULE #14

1. In order for molecules to react, they must collide with each other so that they are close enough to exchange or transfer electrons.

2. Since rate increases with increasing temperature, the reaction at higher temperature will run faster.

3. In order for reactants to react, they must first collide. The more likely collisions are, the faster the reaction will go. When reactants get more concentrated, the vessel that contains them gets more crowded, increasing the chance for collisions between reactants.

4. The units of rate constant depend on the overall order of the reaction. Thus, since the units are different, the overall order must be different as well. The second reaction will be faster than the first because the rate constant is part of the rate equation. Based on the rate equation, if the rate constant is larger, the rate should be larger.

5. When the order of a chemical reaction is zero with respect to one of its reactants, the reaction rate is independent of the concentration of that reactant. Therefore, doubling the concentration of that reactant will not affect the rate at all. This means that the rate will stay the same.

6. The student is not correct. Although temperature is not directly in the rate equation, the value for the rate constant changes with temperature. Thus, since the rate constant is in the rate equation, and since the rate constant depends on temperature, the rate must also depend on temperature.

7. Chemicals that increase reaction rate without getting used up in the process are called catalysts.

8. The higher the activation energy, the lower the reaction rate. Thus, reaction #2 is the fastest.

9. Heterogeneous catalysts are in a different phase from any of the reactants in the chemical equation. Homogeneous catalysts, however, are in the same phase as the reactants. This is the main difference between them. They also use different means to reduce the activation energy of a reaction, so that is an acceptable answer as well.

10. A catalytic converter speeds up the conversion of poisonous carbon monoxide from car exhaust into nonpoisonous carbon dioxide.

ANSWERS TO THE REVIEW QUESTIONS FOR MODULE #15

1. Chemical equilibrium is the point at which the rate of the forward reaction equals the rate of the reverse reaction.

2. Chemical reactions never stop, even when equilibrium is reached. Even though the concentrations of the substances involved do not change, the forward and reverse reactions are still going.

3. The larger the equilibrium constant, the more weighted the reaction is towards products. Thus, the first one will make more products.

4. The student is wrong because all chemical reactions are, in fact, equilibria. We just write those reactions whose K's are enormous as a reaction with only one direction, because there are virtually no reactants left when those reactions reach equilibrium.

5. The concentrations of solids remain constant despite what happens in a chemical reaction. Make sure you understand that *concentration* and *amount* are two different things. Even if I change the amount of a solid, the concentration remains constant.

6. They are important because they determine the strength of an acid.

7. The pH scale ranges from 0 to 14.

8. Solution C is the acidic one. Any solution with pH below 7 is acidic.

9. All of these solutions are acids, so we are really just asking which acid is the weakest acid (low ionization constant means weak acid). The higher the pH, the weaker the acid. Thus, solution B contains the acid with the lowest ionization constant.

10. Acid rain comes from pollutants like sulfur trioxide and nitrogen dioxide that react with water to form acids.

ANSWERS TO THE REVIEW QUESTIONS FOR MODULE #16

1. The charge that an atom in a molecule would develop if the most electronegative atoms in the molecule took the shared electrons from the less electronegative atoms.

2. Since the oxidation number tells us the charge that the atom would have if the more electronegative atoms took the shared electrons from the less electronegative atoms, we can conclude that the atom in question was more electronegative than the other atoms in the first molecule, because a negative charge indicates that it took electrons. The positive oxidation number in the second molecule, however, tells us that it lost electrons to the other atoms in that molecule, indicating that it was less electronegative. Thus, the electronegativities of the other atoms in molecule #1 were lower than the atom in question, while the electronegativities of the atoms in molecule #2 were greater.

3. The sum of all oxidation numbers must equal the charge of the molecule.

4. Oxidation - The process by which an atom loses electrons
Reduction - The process by which an atom gains electrons

5. No, oxidation can never happen without reduction, because a substance cannot lose an electron unless there is something around that can gain it.

6. According to the definition, the anode is where oxidation occurs. In oxidation, a substance loses (gives up) electrons. Thus, the anode is a *source* of electrons. Reduction occurs at the cathode. This means that electrons flow to the cathode so that the substance there can take them. Electrons, then, flow from the anode to the cathode. Thus, anodes are negative and cathodes are positive.

7. If all else fails, assume that the atom's oxidation number is the same as what it would take on in an ionic compound. The atoms that are most likely to follow this rule are in groups 3A, 6A, and 7A. This must be used as a last resort because there are many exceptions to it.

8. A lead-acid battery requires PbO_2, Pb, and H_2SO_4.

9. The internal structure of most batteries gets destroyed as the reaction proceeds. There is no way to reverse the destruction, so the battery is not rechargeable.

10. The alkaline cell has no aqueous solutions, but the lead-acid battery does. The lead-acid battery can be recharged; the alkaline cell cannot be recharged.

SOLUTIONS TO THE PRACTICE PROBLEMS FOR MODULE #1

1. $\frac{1.2 \not{mL}}{1} \times \frac{0.001 \text{ L}}{1 \not{mL}} = \underline{0.0012 \text{ L}}$. The integers in the fractions are exact, and since the "0.001" comes from a definition, it is also exact. Thus, the only number that we need to worry about when it comes to significant figures is the original measurement. It has two significant figures, so our answer can have only two significant figures.

2. $\frac{34.50 \not{km}}{1} \times \frac{1{,}000 \text{ m}}{1 \not{km}} = 34{,}500 \text{ m} = \underline{3.450 \times 10^4 \text{ m}}$. The only number whose significant figures we need to worry about here is the original measurement. All of the other numbers are exact. Since the original measurement has four significant figures, our answer must have four. The only way to express the answer with four significant figures is to use scientific notation. The decimal place in the scientific notation allows us to make one of the zeros significant. 34,500 has only three significant figures, but 3.450 x 10^4 has four significant figures.

3. This is a two-step conversion, since we know of no relationship between km and cm. Thus, we must first convert km to m and then convert m to cm. We'll do this on one line:

$$\frac{0.045 \not{km}}{1} \times \frac{1{,}000 \not{m}}{1 \not{km}} \times \frac{1 \text{ cm}}{0.01 \not{m}} = \underline{4{,}500 \text{ cm}}$$

Since the numbers in the conversion relationships are exact, the only number whose significant figures matter is 0.045. It has two significant figures, so the answer must have two. You could also express your answer as 4.5 x 10^3 cm, as it would have two significant figures as well .

4. This is another two-step conversion. We must convert mL to L and then L to kL, since there is no direct relationship between mL and kL:

$$\frac{34.6 \not{mL}}{1} \times \frac{0.001 \not{L}}{1 \not{mL}} \times \frac{1 \text{ kL}}{1{,}000 \not{L}} = \underline{0.0000346 \text{ kL}}$$

The numbers in the conversion relationships are exact, so the only thing that limits the significant figures is the original measurement. There are three significant figures in the original measurement, so the answer must have three as well. You could also express the answer as 3.46 x 10^{-5} kL; it does not matter.

5. The volume of a box is length times width times height:

$$V = 2.3 \text{ m} \cdot 4.2 \text{ m} \cdot 3.5 \text{ m} = 34 \text{ m}^3$$

All of the numbers have two significant figures, so there should be two in the final answer as well. Unfortunately, the mean guy who wrote the problem wants the answer in cubic centimeters (cc's), so we still have to make a conversion:

$$\frac{34\ m^3}{1} \times \left(\frac{1\ cm}{0.01\ m}\right)^3 = \frac{34\ \cancel{m^3}}{1} \times \frac{1\ cm^3}{0.000001\ \cancel{m^3}} = \underline{3.4 \times 10^7\ cm^3}$$

Notice that to cancel the m^3 unit, I need to cube the conversion relationship between cm and m. There are two significant figures in 34 m^3, and the conversion relationship is exact, so there can only be two significant figures in the answer. The answer could also be expressed as 34,000,000 cm^3.

6. This is a simple problem if you recognize that a cubic centimeter, or cc, is the same thing as a mL:

$$\frac{34.5\ \cancel{mL}}{1} \times \frac{0.001\ L}{1\ \cancel{mL}} = \underline{0.0345\ L}$$

The numbers in the conversion relationship are exact, so the only significant figures that matter are in 34.5 mL. That number has three significant figures, so the answer can have only three.

7. a. This is a big number, so the exponent needs to be positive. To get the decimal next to the first digit, we have to move it two places. The answer, then, is $\underline{1.2345 \times 10^2}$.

   b. This is a small number, so the exponent has to be negative. The decimal point needs to be moved four places to get to the right of the first digit. The answer, then, is $\underline{3.040 \times 10^{-4}}$. Note that the last zero *must* be there, as it is significant in the original number.

   c. This is a big number, so the exponent is positive, and the decimal must be moved six places to get it next to the first digit. Thus, the answer is $\underline{6.1 \times 10^6}$. Note that I dropped the zeros in the original number, because none of them are significant.

   d. This number is small, so the exponent is negative, and the decimal need be moved only one place. The answer, then, is $\underline{1.234 \times 10^{-1}}$.

8. a. A positive exponent tells us to make this a big number by moving the decimal 3 places, so the answer is 6,540. The zero is not significant, so this number has the same significant figures as the original number.

   b. A negative exponent means to make the number small by moving the decimal 3 places, so the answer is 0.003450. The last zero must be there, as it is significant in the original number.

   c. A positive exponent tells us to make this a big number by moving the decimal 7 places, so the answer is 35,600,000. The zeros are not significant, so this number has the same significant figures as the original number.

   d. A negative exponent tells us to make the number small by moving the decimal 7 places, so the answer is 0.0000004050. The last zero must be there, as it is significant in the original number.

9. We need to use Equation (1.1) here, but we must do two things. First, we have to get the units to agree. The density is given in grams per mL, but the volume is in L. First, then, we have to convert the volume to mL:

$$\frac{3.45\ \cancel{L}}{1} \times \frac{1\ mL}{0.001\ \cancel{L}} = 3.45 \times 10^3\ mL$$

Now that the units work out, we must also rearrange Equation (1.1) to solve for mass:

$$m = \rho \cdot V$$

$$m = 11.4\ \frac{g}{\cancel{mL}} \cdot 3.45 \times 10^3\ \cancel{mL}$$

$$m = \underline{3.93 \times 10^4\ g}$$

Since each number in the equation has three significant figures, the answer must have three.

10. This problem is like #9, but first we have to convert kg to g to make our units consistent:

$$\frac{45.6\ \cancel{kg}}{1} \times \frac{1{,}000\ g}{1\ \cancel{kg}} = 4.56 \times 10^4\ g$$

Now we can rearrange Equation (1.1) to solve for volume:

$$V = \frac{m}{\rho}$$

$$V = \frac{4.56 \times 10^4\ g}{19.3\ \frac{g}{cc}}$$

$$V = \underline{2.36 \times 10^3\ cc}$$

Since the problem did not request any specific units, we can simply leave our answer in cc's. Also, since each number in the equation has three significant figures, the answer must have three.

SAMPLE CALCULATIONS FOR EXPERIMENT 1.3

Length of the book: 10.95 in **Converted to cm**: $\frac{10.95 \text{ \cancel{in}}}{1} \times \frac{2.54 \text{ cm}}{1 \text{ \cancel{in}}} = 27.81 \text{ cm}$

I can have four significant figures because 2.54 cm is exact, as I stated in the instructions. Thus, the only number to consider in this conversion is 10.95 in.

Width of the book: 8.54 in **Converted to cm**: $\frac{8.54 \text{ \cancel{in}}}{1} \times \frac{2.54 \text{ cm}}{1 \text{ \cancel{in}}} = 21.7 \text{ cm}$

Surface area of the book: (10.95 in)·(8.54 in) = 93.5 in^2
There are three significant figures in the answer because 8.54 has only three significant figures.

Measured length of the book in cm: 27.76 cm (Very close to the converted value listed above.)

Measured width of the book in cm: 21.56 cm (Not too far off the converted value listed above.)

Surface area of the book: (27.76 cm)·(21.56 cm) = 598.5 cm^2

Surface area converted to in^2:

$$\frac{598.5 \text{ cm}^2}{1} \times \left(\frac{1 \text{ in}}{2.54 \text{ cm}}\right)^2 = \frac{598.5 \text{ \cancel{cm}}^2}{1} \times \frac{1 \text{ in}^2}{6.4516 \text{ \cancel{cm}}^2} = 92.77 \text{ in}^2$$

Note that I have four significant figures because the conversion relationship is exact. This value is pretty close to the measured value for the surface area in inches.

SAMPLE CALCULATIONS FOR EXPERIMENT 1.4

Mass of the graduated cylinder: 25 g
Mass of the syrup and the cylinder: 82 g
You can read your mass scale to 1 g, since it is probably marked off in increments of 10 g.

Mass of the syrup: 82 g - 25 g = 57 g
Since both masses have a significant figure in the ones place, the answer can be reported to the ones place.

Density of the syrup: $\frac{57 \text{ g}}{50.0 \text{ mL}} = 1.1 \frac{\text{g}}{\text{mL}}$

Your density might be higher, depending on the syrup used. The answer has two significant figures because 57 has two significant figures.

Mass of the water and the cylinder: 75 g

Mass of the water: 75 g - 25 g = 5.0×10^1 g
Since both masses have a significant figure in the ones place, the answer can be reported to the ones place. In this case, I had to use scientific notation to indicate that the zero is significant.

Density of the water: $\frac{5.0 \times 10^1 \text{ g}}{50.0 \text{ mL}} = 1.0 \frac{\text{g}}{\text{mL}}$

The answer has two significant figures because 5.0×10^1 has two significant figures.

Mass of the oil and the cylinder: 71 g

Mass of the oil: 71 g - 25 g = 46 g
Since both masses have a significant figure in the ones place, the answer can be reported to the ones place.

Density of the oil: $\frac{46 \text{ g}}{50.0 \text{ mL}} = 0.92 \frac{\text{g}}{\text{mL}}$

The answer has two significant figures because 46 has two significant figures.

SOLUTIONS TO THE PRACTICE PROBLEMS FOR MODULE #2

1. We solve this problem by rearranging Equation (2.1):

$$^{\circ}F = \frac{9}{5}\cdot\left(^{\circ}C\right) + 32$$

$$^{\circ}F = \frac{9}{5}\cdot(15.0) + 32 = 59.0$$

So 15.0 °C is the same as 59.0 °F. Note that since the 9, 5, and 32 in this equation are exact, the only significant figures we need to count are in the original measurement. Since 15.0 °C has three significant figures, our answer must have three.

2. First we convert from K to °C by using a rearranged form of Equation (2.2):

$$^{\circ}C = K - 273.15$$

$$^{\circ}C = 3.5 - 273.15 = -269.65$$

Since we are subtracting here, we must follow the rule for addition and subtraction. The number 3.5 has its last significant figure in the tenths place, while 273.15 has its last significant figure in the hundredths place. Thus, we can report our answer only to the tenths place. This means that the Celsius temperature is -269.7 °C.

Now we can convert to Fahrenheit:

$$^{\circ}F = \frac{9}{5}\cdot\left(^{\circ}C\right)+32$$

$$^{\circ}F = \frac{9}{5}\cdot(-269.7)+32 = -453.5$$

Since our Celsius temperature had four significant figures, and since the other numbers in the equation are exact, the Fahrenheit temperature must have four significant figures. Thus, the answer is -453.5 °F.

3. This problem is a direct application of Equation (2.1):

$$^{\circ}C = \frac{5}{9}\cdot\left(^{\circ}F-32\right)$$

$$^{\circ}C = \frac{5}{9}\cdot\left(115-32\right) = 46.1$$

Since all of the numbers except 115 are exact, the answer must have the same number of significant figures as 115 has. A hot temperature of 115 °F at least sounds cooler when you say 46.1 °C.

4. One food Calorie = 1,000 calories. We must use this fact first, because the only relationship that we know which contains Joules is the one that relates Joules to chemistry calories. So we first have to convert food calories to chemistry calories:

$$\frac{2{,}500.0\ \cancel{\text{Cal}}}{1} \times \frac{1{,}000\ \text{cal}}{1\ \cancel{\text{Cal}}} = 2.5000 \times 10^{6}\ \text{cal}$$

Since the conversion relationship is exact, the answer must have the same number of significant figures as the original measurement. Thus, the answer must have five significant figures. Now we can convert calories to Joules with the relationship 1 cal = 4.184 J:

$$\frac{2.5000 \times 10^{6}\ \cancel{\text{cal}}}{1} \times \frac{4.184\ \text{J}}{1\ \cancel{\text{cal}}} = 1.046 \times 10^{7}\ \text{J}$$

So the average person burns $\underline{1.046 \text{ x } 10^{7} \text{ Joules}}$ of energy per day! Remember, this conversion relationship is not exact. The integers, of course, are still exact, but as I mentioned in the module, the 4.184 is not exact. Since it contains only four significant figures, the answer can have only four significant figures.

5. According to Table 2.1, glass has a specific heat of $0.8372\ \frac{\text{J}}{\text{g}\cdot{}^{\text{o}}\text{C}}$. Since we have specific heat, mass, initial temperature, and final temperature, we are obviously supposed to use Equation (2.3). Before we can do that, though, we have to get our mass in grams so that its units agree with the specific heat units:

$$\frac{15.1\ \cancel{\text{kg}}}{1} \times \frac{1{,}000\ \text{g}}{1\ \cancel{\text{kg}}} = 1.51 \times 10^{4}\ \text{g}$$

Now we can use Equation (2.3):

$$q = m \cdot c \cdot \Delta T$$

$$q = \left(1.51 \times 10^{4}\ \text{g}\right) \cdot \left(0.8372 \frac{\text{J}}{\text{g}\cdot{}^{\text{o}}\text{C}}\right) \cdot \left(45\ {}^{\text{o}}\text{C} - 15\ {}^{\text{o}}\text{C}\right)$$

$$q = \left(1.51 \times 10^{4}\ \cancel{\text{g}}\right) \cdot \left(0.8372 \frac{\text{J}}{\cancel{\text{g}}\cdot{}^{\text{o}}\cancel{\text{C}}}\right) \cdot \left(3.0 \times 10^{1}\ {}^{\text{o}}\cancel{\text{C}}\right) = \underline{3.8 \times 10^{5}\ \text{J}}$$

Now look at the significant figures in this problem. First, we subtracted 15 from 45. For that, we had to use the rule of addition and subtraction. Since both 45 and 15 have their last significant figure in the ones place, the answer must have its last significant figure in the ones place. Thus, the only way I could report that answer to indicate that the resulting zero in the ones place is significant is to use scientific notation. That's why the change in temperature is reported at $3.0 \text{ x } 10^{1}\ {}^{\text{o}}\text{C}$. After that, the equation uses only multiplication, so at that point, we must count significant figures. The lowest number of significant figures in the equation is two ($3.0 \text{ x } 10^{1}\ {}^{\text{o}}\text{C}$), so the answer can have only two.

6. In this problem, we are given ΔT, mass, and the heat absorbed, and we must calculate specific heat. So, first we rearrange Equation (2.3) to solve for specific heat, and then we plug our numbers in:

$$c = \frac{q}{m \cdot \Delta T}$$

$$c = \frac{50.0 \text{ kJ}}{(124.1 \text{ g}) \cdot (36.3 \text{ °C})} = \underline{0.0111 \frac{\text{kJ}}{\text{g} \cdot \text{°C}}}$$

You might wonder why I didn't convert kJ into J. Well, in this problem, I was given no restrictions on units. I didn't need any units to cancel out, and the problem didn't specify what units to give specific heat in; thus, I didn't need to convert the energy unit. This unit is a perfectly acceptable unit for specific heat. If you did convert from kJ to J, your answer should be $11.1 \frac{\text{J}}{\text{g} \cdot \text{°C}}$.

7. In order to get the copper's new temperature, we need to solve for ΔT in Equation (2.3). We can do this because we have the mass and heat given in the problem and the specific heat from Table 2.1. Remember, though, since the copper lost heat, its q is negative! So first we rearrange the equation to solve for ΔT:

$$\Delta T = \frac{q}{m \cdot c}$$

$$\Delta T = \frac{-456.7 \cancel{\text{J}}}{(245 \cancel{\text{g}}) \cdot \left(0.3851 \frac{\cancel{\text{J}}}{\cancel{\text{g}} \cdot \text{°C}}\right)} = -4.84 \text{ °C}$$

Now that we have ΔT, we can rearrange Equation (2.4) to solve for final temperature:

$$T_{final} = \Delta T + T_{initial}$$

$$T_{final} = -4.84 \text{ °C} + 25 \text{ °C} = \underline{2.0 \times 10^{1} \text{ °C}}$$

Note that the final temperature is lower than the initial temperature, which should make sense since the copper lost energy. Also, notice that according to the rule of addition and subtraction, I must report my answer to the ones place. That means the resulting zero in the ones place is significant. The only way you can report this answer properly, then, is with scientific notation.

8. You are supposed to have the specific heat of water memorized. Thus, you have all the information needed to calculate the heat required to raise water from 0.0 °C to 37.0 °C:

$$q = m \cdot c \cdot \Delta T$$

$$q = (3.40 \times 10^2 \text{ g}) \cdot \left(1.000 \frac{\text{cal}}{\text{g} \cdot {}^\circ\text{C}}\right) \cdot (37.0 \ {}^\circ\text{C} - 0.0 \ {}^\circ\text{C})$$

$$q = (3.40 \times 10^2 \text{ g}) \cdot \left(1.000 \frac{\text{cal}}{\text{g} \cdot {}^\circ\text{C}}\right) \cdot (37.0 \ {}^\circ\text{C}) = 1.26 \times 10^4 \text{ cal}$$

After subtracting 0.0 from 37.0, we are left with a ΔT of 37.0 °C. Since both 37.0 and 3.40 x 10^2 each contain three significant figures, the answer can have only three. That's a lot of calories, but the problem asks for the answer in Calories (with a capital "C"), which means food calories. It takes 1,000 cal to make 1 Cal:

$$\frac{1.26 \times 10^4 \text{ cal}}{1} \times \frac{1 \text{ Cal}}{1{,}000 \text{ cal}} = 12.6 \text{ Cal}$$

Drinking a 12-ounce glass of ice-cold water burns 12.6 Calories when you drink it.

9. This is a standard calorimetry problem such as the one in the second part of Example 2.4. We must therefore use the calorimetry equation. To use that equation, however, we need to calculate as many q's as we can. We know that $q_{calorimeter} = 0$, since the problem says that we can ignore the calorimeter. We also have enough information to calculate the heat absorbed by the water:

$$q_{water} = m \cdot c \cdot \Delta T$$

$$q_{water} = (150.0 \text{g}) \cdot \left(4.184 \frac{\text{J}}{\text{g} \cdot {}^\circ\text{C}}\right) \cdot (5.4 \ {}^\circ\text{C})$$

$$q_{water} = 3.4 \times 10^3 \text{ J}$$

Now that we know q_{water} and $q_{calorimeter}$, we can use the calorimetry equation to determine the heat lost by the metal:

$$-q_{object} = q_{water} + q_{calorimeter}$$

$$-q_{object} = 3.4 \text{ x } 10^3 \text{ J}$$

We can use this information in Equation (2.3) to determine the specific heat of the metal. However, we have to determine the ΔT of the metal. The metal starts at 100.0 °C and ends at the same temperature as the water and calorimeter. Well, if the water started out at 24.1 °C, and its temperature raised 5.4 °C, that tells us that the final temperature of the water (which is the same as the final temperature of the metal) must be 29.5 °C.

$$c = \frac{q}{m \cdot \Delta T} = \frac{-3.4 \times 10^3 \text{ J}}{(50.0 \text{g}) \cdot (29.5 \ {}^\circ\text{C} - 100.0 \ {}^\circ\text{C})} = 0.96 \frac{\text{J}}{\text{g} \cdot {}^\circ\text{C}}$$

10. When you look at this problem, you see that you're trying to discover the specific heat of the calorimeter, not the specific heat of the metal, as is usually the case. Instead, the specific heat of the metal can be found in Table 2.1 ($c = 0.3851 \frac{J}{g \cdot {}^{o}C}$). In order to find the calorimeter's specific heat, we will need to use the calorimetry equation. Thus, we need to figure out as many q's as we can. We have enough information to determine the q's of the metal and the water, so we should start there:

$$q_{metal} = m \cdot c \cdot \Delta T$$

$$q_{metal} = (345.1\ g) \cdot \left(0.3851 \frac{J}{g \cdot {}^{o}C}\right) \cdot (25.1\ {}^{\circ}C - 100.0\ {}^{\circ}C)$$

$$q_{metal} = (345.1\ \cancel{g}) \cdot \left(0.3851 \frac{J}{\cancel{g} \cdot {}^{o}\cancel{C}}\right) \cdot (-74.9\ {}^{\circ}\cancel{C}) = -9{,}950\ J$$

$$q_{water} = m \cdot c \cdot \Delta T$$

$$q_{water} = (150.0\,g) \cdot \left(4.184 \frac{J}{g \cdot {}^{o}C}\right) \cdot (25.1\ {}^{\circ}C - 24.2\ {}^{\circ}C)$$

$$q_{water} = (150.0\,\cancel{g}) \cdot \left(4.184 \frac{J}{\cancel{g} \cdot {}^{o}\cancel{C}}\right) \cdot (0.9\ {}^{\circ}\cancel{C}) = 600\ J$$

Notice what happened with the significant figures when we calculated q_{water}. In determining ΔT, we can only report our answer to the tenths place, since both temperature measurements have their last significant figure in the tenths place. This leaves us with only one significant figure for ΔT. Thus, when we multiply by ΔT, the answer can have only one significant figure and therefore must be rounded up to 600 J. We can now use the calorimetry equation:

$$-q_{object} = q_{water} + q_{calorimeter}$$

$$-(-9950\ J) = 600\ J + q_{calorimeter}$$

$$q_{calorimeter} = 9{,}400$$

Once again, look at the significant figures. To get the answer, we subtract 600 from 9950. Since the least precise number (600) has its only significant figure in the hundreds place, the answer must be given to the hundreds place. We can now use the heat of the calorimeter in Equation (2.3) to determine the specific heat of the calorimeter:

$$c = \frac{q}{m \cdot \Delta T} = \frac{9400\ J}{(4.5\ g) \cdot (0.9\ {}^{\circ}C)} = \underline{2{,}000 \frac{J}{g \cdot {}^{o}C}}$$

Once again, consider the significant figures. As discussed above, the value for ΔT can be reported only to the tenths place. This leaves only one significant figure, so the answer must be rounded down to 2,000 $\frac{J}{g \cdot {}^{o}C}$.

SAMPLE CALCULATIONS FOR EXPERIMENT 2.2

Mass of the metal: 53 g
Mass of the calorimeter: 5 g
Mass of the calorimeter + water: 79 g
Mass of the water: 79 g - 5 g = 74 g
We can report the mass to the ones place, since this is subtraction and both measurements used in the subtraction are reported to the ones place.

Initial temperature: 24.8 °C
Most likely, your thermometer is marked off in individual degrees (or in units of 2 degrees). You are supposed to estimate to the next decimal place, so you should report your temperatures to the tenths place.

Final temperature: 27.4 °C

$$\Delta T = 27.4\ {}^{o}C - 24.8\ {}^{o}C = 2.6\ {}^{o}C$$

Since both temperatures are reported to the tenths place, the difference must be reported to the tenths place.

$$q_{water} = m \cdot c \cdot \Delta T$$

$$q_{water} = (74\ g) \cdot \left(4.184\ \frac{J}{g \cdot {}^{o}C}\right) \cdot (2.6\ {}^{o}C) = 810\ J$$

Since 74 and 2.6 have only two significant figures, the result of this multiplication can have only two significant figures. Since we are ignoring the calorimeter, we can easily calculate the heat lost by the metal:

$$-q_{object} = q_{water} + q_{calorimeter}$$

$$q_{object} = -810\ J$$

Now we can determine the specific heat:

$$c = \frac{q}{m \cdot \Delta T} = \frac{-810\ J}{(53\ g) \cdot (27.4\ {}^{o}C - 100.0\ {}^{o}C)} = 0.21\ \frac{J}{g \cdot {}^{o}C}$$

There can be only two significant figures in the specific heat, because the mass and heat have only two.

SOLUTIONS TO THE PRACTICE PROBLEMS FOR MODULE #3

1. Since the chemist starts with 150.0 g of matter, he has to end up with 150.0 g of matter. Since hydrogen and oxygen are the only two elements in water, any mass not in the hydrogen must be in the oxygen:

$$\text{Total mass} = \text{Mass of hydrogen} + \text{Mass of oxygen}$$

$$\text{Mass of oxygen} = \text{Total mass} - \text{Mass of hydrogen}$$

$$\text{Mass of oxygen} = 150.0\text{ g} - 16.7\text{ g} = 133.3\text{ g}$$

So the chemist must have made 133.3 g of oxygen during the experiment.

2. Since the compound has only carbon, nitrogen, and hydrogen in it, the sum of those individual masses must add up to the mass of the compound:

$$\text{Mass of compound} = \text{Mass of nitrogen} + \text{Mass of carbon} + \text{Mass of hydrogen}$$

$$\text{Mass of compound} = 14.0\text{ g} + 12.0\text{ g} + 3.0\text{ g} = 29.0\text{ g}$$

By the law of mass conservation, then, the mass of the compound before the decomposition must have been 29.0 g.

3. All elements that lie to the left of the jagged line on the chart are metals, while all elements to the right of the jagged line are non-metals. Thus, Y and Re are metals while I and N are nonmetals.

4. The chemist starts with 20.0 g + 20.0 g = 40.0 g of matter; thus, she must end up with 40.0 g of matter. According to the problem, she made 21.3 g of hydrogen peroxide along with left over hydrogen. Since all 40.0 g must be accounted for, the remaining mass must be in the hydrogen:

$$\text{Mass of hydrogen} = \text{Total mass} - \text{Mass of product}$$

$$\text{Mass of hydrogen} = 40.0\text{ g} - 21.3\text{ g} = 18.7\text{ g}$$

By the law of mass conservation, then, there were 18.7 g of hydrogen left over. Since we started out with 20.0 g of hydrogen and there were 18.7 g left over, only 20.0 g -18.7 g = 1.3 g were actually used to make hydrogen peroxide. The correct recipe, then, is 1.3 g of hydrogen for every 20.0 g of oxygen.

5. The reaction starts with 100.0 g + 100.0 g = 200.0 g of matter; thus, there must be 200.0 g of matter after everything is finished. According to the problem, these amounts of calcium and nitrogen made 152.5 g of calcium nitride along with left over nitrogen. Since all 200.0 g must be accounted for, the remaining mass must be in the nitrogen:

$$\text{Mass of nitrogen} = \text{Total mass} - \text{Mass of product}$$

$$\text{Mass of nitrogen} = 200.0 \text{ g} - 152.5 \text{ g} = 47.5 \text{ g}$$

By the law of mass conservation, then, there were 47.5 g of nitrogen left over. Since we started out with 100.0 g of nitrogen and there were 47.5 g left over, only 100.0 g - 47.5 g = 52.5 g were actually used to make calcium nitride. Thus, the proper recipe for making 152.5 g of calcium nitride is to add 100.0 g of calcium to 52.5 g of nitrogen.

The problem, however, asks us the recipe for making 1.0 kg, or 1.0×10^3 g of calcium nitride. Therefore, we need to determine how much to increase the amount of ingredients in order to make this larger amount:

$$(152.5 \text{ g}) \cdot x = 1.0 \times 10^3 \text{ g}$$

$$x = \frac{1.0 \times 10^3 \text{ g}}{152.5 \text{ g}} = 6.6$$

To make 1.0 kg, then, we just multiply the amount of each component by 6.6:

$$\text{Mass of calcium} = 100.0 \text{ g} \cdot 6.6 = 6.6 \times 10^2 \text{ g}$$

$$\text{Mass of nitrogen} = 52.5 \text{ g} \cdot 6.6 = 3.5 \times 10^2 \text{ g}$$

You need 6.6×10^2 g of calcium and 3.5×10^2 g of nitrogen to make 1.0 kg of calcium nitride.

6. Put the number of atoms as a subscript *after* each element symbol. If the number is a "1," drop it: $C_3H_6Cl_2O$

7. To get the total number of atoms in the molecule, we have to realize that each subscript tells us how many of each atom it takes to make the molecule. Thus, the total number of atoms would just be the sum of the subscripts, remembering that if there is no number, we assume it is "1." There are, therefore, 48 atoms in one molecule of this compound.

8. Only SO_3 and P_2H_6 are made up entirely of nonmetals, so (a) and (b) are covalent.

9. a. This is a covalent compound, so we have to use prefixes. The prefix for sulfur is "mono," but we drop it because sulfur is the first molecule. The prefix for oxygen is "tri," and we change oxygen's ending to "ide." The name is sulfur trioxide

   b. This is a covalent compound so we have to use prefixes. The prefix for phosphorus is "di." The prefix for hydrogen is "hexa," so the name is diphosphorus hexahydride

   d. This is an ionic compound so we don't use prefixes: potassium fluoride

10. a. The prefix "tetra" means "four" and the prefix "hexa" means 6. The formula is N_4H_6.
    b. No prefix on the first atom means "one," and the "mono" prefix also means "one." The formula is NO.

SOLUTIONS TO THE PRACTICE PROBLEMS FOR MODULE #4

1. a. An egg is a mixture because it has a shell, yolk, and white, which are all different substances.
   b. A gold nugget is pure gold, so it is a pure substance.
   c. Since the bottle has only HNO_3 in it, it has only one substance and is therefore a pure substance.
   d. Lemonade dissolved in water is a mixture, because the lemonade retains its properties, as does the water.

2. a. An egg is a heterogeneous mixture because the shell, yolk, and white are all in separate places.
   b. Gold (Au) is an element.
   c. HNO_3 is a compound, because it is a molecule made up of more than one element.
   d. Lemonade dissolved in water is a homogenous mixture because its composition is the same throughout the sample.

3. a. This is a physical change because the vase's molecules did not change. It is also easily reversible: The vase could be put back together.
   b. This is a chemical change. The molecules in the different ingredients actually react when they are baked to form new molecules. The other way to look at this is you cannot "unbake" a cake.
   c. This is a physical change, because the CO_2 is just changing phase.
   d. This is a chemical change. Burning charcoal turns the carbon into CO_2. The other way to look at it is that you cannot "unburn" charcoal.
   e. This is a physical change. Dissolving one substance into another never changes their molecules. Alternatively, you could easily reverse this by boiling off the water to recover the Kool-Aid.

4. Nitrogen is a homonuclear diatomic, so its formula is N_2.

5.

Reactants Side	Products Side	
#Ca: 1x1 = 1	#Ca 1x1 = 1	Balanced with respect to Ca.
#F: 1x2 = 2	#F: 1x1 = 1	NOT balanced with respect to F
#N: 2x1 = 2	#N 1x1 = 1	NOT balanced with respect to N
#H: 2x4 = 8	#H: 1x4 = 4	NOT balanced with respect to H
#Cl: 2x1 = 2	#Cl: 1x2 = 2	Balanced with respect to Cl

The equation is not balanced.

6. To balance, we must first count up the atoms:

Reactants Side	Products Side	
#H: 1x1 = 1	#H: 1x2 = 2	NOT balanced with respect to H
#Cl: 1x1 = 1	#Cl: 1x2 = 2	NOT balanced with respect to Cl
#Zn: 1x1 = 1	#Zn: 1x1 = 1	Balanced with respect to Zn

We can start by balancing the H's. There are twice as many on the products side as the reactants side, so we need to double the H's on the reactants side:

$$\underline{2HCl\ (aq) + Zn\ (s) \rightarrow ZnCl_2\ (aq) + H_2\ (g)}$$

Counting atoms now gives us:

Reactants Side	Products Side	
#H: 2x1 = 2	#H: 1x2 = 2	Balanced with respect to H
#Cl: 2x1 = 2	#Cl: 1x2 = 2	Balanced with respect to Cl
#Zn: 1x1 = 1	#Zn: 1x1 = 1	Balanced with respect to Zn

The equation is now balanced.

7. Turning the words into equation form:

$$N_2 + O_2 \rightarrow NO$$

Counting atoms:

Reactants Side	Products Side	
#N: 1x2 = 2	#N: 1x1 = 1	NOT balanced with respect to N
#O: 1x2 = 2	#O: 1x1 = 1	NOT balanced with respect to O

To balance the N's we must double the number of N's on the products side:

$$\underline{N_2 + O_2 \rightarrow 2NO}$$

Counting atoms again:

Reactants Side	Products Side	
#N: 1x2 = 2	#N: 2x1 = 2	Balanced with respect to N
#O: 1x2 = 2	#O: 2x1 = 2	Balanced with respect to O

The equation is now balanced.

8. Turning the words into equation form:

$$C_7H_{10}\,(l) + H_2\,(g) \rightarrow C_7H_{16}\,(l)$$

Counting atoms:

Reactants Side	Products Side	
#C: 1x7 = 7	#C: 1x7 = 7	Balanced with respect to C
#H: 1x10 + 1x2=12	#H: 1x16 = 16	NOT balanced with respect to H

To balance this equation, we must add 6 more H's to the reactants side. To do this, we had better not mess with the C_7H_{10}, because that will ruin the balance we have for carbon. Thus, we should change the number next to the H_2. Since 10 H's come from the C_7H_{10}, we need 6 H's to come from H_2. Therefore:

$$\underline{C_7H_{10}\,(l) + 3H_2\,(g) \rightarrow C_7H_{16}\,(l)}$$

Counting atoms again:

Reactants Side	Products Side	
#C: 1x7 = 7	#C: 1x7 = 7	Balanced with respect to C
#H: 1x10 + 3x2=16	#H: 1x16 = 16	Balanced with respect to H

9. Counting atoms:

Reactants Side	Products Side	
#C: 1x7 = 7	#C: 1x1 = 1	NOT balanced with respect to C
#H: 1x16=16	#H: 1x2 = 2	NOT balanced with respect to H
#O: 1x2 = 2	#O: 1x2+1x1 = 3	NOT balanced with respect to O

Balancing the C's and H's at the same time:

$$C_7H_{16}\ (l)\ +\ O_2\ (g)\ \rightarrow\ 7CO_2\ (g)\ +\ 8H_2O\ (g)$$

Counting atoms:

Reactants Side	Products Side	
#C: 1x7 = 7	#C: 7x1 = 7	Balanced with respect to C
#H: 1x16=16	#H: 8x2 = 16	Balanced with respect to H
#O: 1x2 = 2	#O: 7x2+8x1 = 22	NOT balanced with respect to O

To balance the O's, we must multiply the O's on the reactants side by 11:

$$\underline{C_7H_{16}\ (l)\ +\ 11O_2\ (g)\ \rightarrow\ 7CO_2\ (g)\ +\ 8H_2O\ (g)}$$

Counting atoms:

Reactants Side	Products Side	
#C: 1x7 = 7	#C: 7x1 = 7	Balanced with respect to C
#H: 1x16=16	#H: 8x2 = 16	Balanced with respect to H
#O: 11x2 = 22	#O: 7x2+8x1 = 22	Balanced with respect to O

10. Counting atoms:

Reactants Side	Products Side	
#C: 1x1 = 1	#C: 1x12 = 12	NOT balanced with respect to C
#H: 1x2 = 2	#H: 1x24 = 24	NOT balanced with respect to H
#O: 1x2 + 1x1 = 3	#O: 1x12 + 1x2 = 14	NOT balanced with respect to O

To balance this, we need to leave the O's until the end, because they appear in all of the substances. We can balance the C's and the H's in one step:

$$12CO_2\ (g)\ +\ 12H_2O\ (l)\ \rightarrow\ C_{12}H_{24}O_{12}\ (s)\ +\ O_2\ (g)$$

Counting atoms again:

Reactants Side	Products Side	
#C: 12x1 = 12	#C: 1x12 = 12	Balanced with respect to C
#H: 12x2=24	#H: 1x24 = 24	Balanced with respect to H
#O: 12x2+12x1=36	#O: 1x12+1x2 = 14	NOT balanced with respect to O

To balance the O's we can't mess with the $C_{12}H_{24}O_{12}$ because changing its number will throw our C's and H's off balance. Thus, we must work with the O_2. Since 12 O's come from the $C_{12}H_{24}O_{12}$, we need 24 more to come from the O_2. Therefore

$$12CO_2\ (g) + 12H_2O\ (l) \rightarrow C_{12}H_{24}O_{12}\ (s) + 12O_2\ (g)$$

Counting atoms:

Reactants Side	Products Side	
#C: 12x1 = 12	#C: 1x12 = 12	Balanced with respect to C
#H: 12x2=24	#H: 1x24 = 24	Balanced with respect to H
#O: 12x2+12x1=36	#O: 1x12+12x2=36	Balanced with respect to O

SOLUTIONS TO THE PRACTICE PROBLEMS FOR MODULE #5

1. The compound has Rb, N, and O in it. Remember, N and O are homonuclear diatomics!

$$RbNO_3 \rightarrow Rb + N_2 + O_2$$

The Rb's are already balanced. To balance the N's and O's without disturbing the Rb's, we will have to use fractions:

$$RbNO_3 \rightarrow Rb + \frac{1}{2}N_2 + \frac{3}{2}O_2$$

Eliminating the fractions:

$$2 \text{ x } [RbNO_3] \rightarrow 2 \text{ x } [Rb + \frac{1}{2}N_2 + \frac{3}{2}O_2]$$

$$\underline{2RbNO_3 \rightarrow 2Rb + N_2 + 3O_2}$$

2. The compound is made up of Na, H, S, and O. H and O are homonuclear diatomics, thus:

$$Na + H_2 + S + O_2 \rightarrow NaHSO_4$$

The Na's and S's are already balanced. To balance the H's without disturbing the others, we will use a fraction. The O's are easy to balance:

$$Na + \frac{1}{2}H_2 + S + 2O_2 \rightarrow NaHSO_4$$

Eliminating the fraction:

$$2 \text{ x } [Na + \frac{1}{2}H_2 + S + 2O_2] \rightarrow 2 \text{ x } [NaHSO_4]$$

$$\underline{2Na + H_2 + 2S + 4O_2 \rightarrow 2NaHSO_4}$$

3. Complete combustion reactions have O_2 and whatever is being burned as reactants while CO_2 and H_2O are products:

$$C_3H_6O + O_2 \rightarrow CO_2 + H_2O$$

Balancing the C's and H's is easy:

$$C_3H_6O + O_2 \rightarrow 3CO_2 + 3H_2O$$

There are 9 O's on the products side and 3 on the reactants side. I better not mess with the C_3H_6O molecule, though, because that would throw off the C and H balance:

$$\underline{C_3H_6O + 4O_2 \rightarrow 3CO_2 + 3H_2O}$$

4. A Sn atom has a mass of 118.7 amu according to the chart. All we have to do is convert this to grams:

$$\frac{118.7\ \text{amu}}{1} \times \frac{1.66 \times 10^{-24}\ \text{g}}{1.00\ \text{amu}} = 1.97 \times 10^{-22}\ \text{g}$$

An Sn atom has a mass of 1.97 x 10^{-22} g.

5. A K_2CrO_4 molecule has 2 K's (39.1 amu each), 1 Cr (52.0 amu), and 4 O's (16.0 amu each), thus:

Mass of K_2CrO_4 = 2 x 39.1 amu + 52.0 amu + 4 x 16.0 amu = 194.2 amu

Remember, since we are adding here, we look at decimal place. Each mass is reported to the tenths place, so the answer can be reported to the tenths place. Since we can only relate amu to g, we need to do a two-step conversion:

$$\frac{194.2\ \text{amu}}{1} \times \frac{1.66 \times 10^{-24}\ \text{g}}{1.00\ \text{amu}} \times \frac{1\ \text{kg}}{1{,}000\ \text{g}} = 3.22 \times 10^{-25}\ \text{kg}$$

A K_2CrO_4 molecule has a mass of 3.22 x 10^{-25} kg.

6. The periodic chart tells us that one Cr atom has a mass of 52.0 amu. Therefore:

52.0 grams Cr = 1 mole Cr

This is the conversion relationship we need to convert grams of Cr into moles of Cr:

$$\frac{200.0\ \text{g Cr}}{1} \times \frac{1\ \text{mole Cr}}{52.0\ \text{g Cr}} = 3.85\ \text{moles Cr}$$

There are 3.85 moles of Cr in this sample.

7. The molecular mass of $NaHCO_3$ is:

1 x 23.0 amu + 1 x 1.01 amu + 1 x 12.0 amu + 3 x 16.0 amu = 84.0 amu

Using the rule of addition and subtraction, we can only report our answer to the tenths place, because the least precise numbers in the problem (22.0, 12.0, and 16.0) have their last significant figure in the tenths place. This means:

84.0 g $NaHCO_3$ = 1 mole $NaHCO_3$

Now we do the grams to moles conversion:

$$\frac{125\ \text{g NaHCO}_3}{1} \times \frac{1\ \text{mole NaHCO}_3}{84.0\ \text{g NaHCO}_3} = 1.49\ \text{moles NaHCO}_3$$

There are 1.49 moles of $NaHCO_3$ in this sample.

8. This is just another conversion problem, but this time we are converting moles into grams. We still need to determine the molecular mass of $CuCl_2$ first, however:

$$\text{Mass of } CuCl_2 = 1 \text{ x } 63.5 \text{ amu} + 2 \text{ x } 35.5 \text{ amu} = 134.5 \text{ amu}$$

This means:

$$134.5 \text{ grams } CuCl_2 = 1 \text{ mole } CuCl_2$$

Now we can do our conversion:

$$\frac{0.172 \cancel{\text{moles} CuCl_2}}{1} \times \frac{134.5 \text{ g} CuCl_2}{1 \cancel{\text{mole} CuCl_2}} = 23.1 \text{ g} CuCl_2$$

This sample has a mass of <u>23.1 g $CuCl_2$</u>.

9. Once again, this is just a conversion problem. We first look at the chart to determine the relationship between grams and moles:

$$140.1 \text{ g Ce} = 1 \text{ mole Ce}$$

Now we use this relationship in a conversion:

$$\frac{15.0 \cancel{\text{moles Ce}}}{1} \times \frac{140.1 \text{ g Ce}}{1 \cancel{\text{mole Ce}}} = 2.10 \times 10^3 \text{ g Ce}$$

This sample has a mass of <u>$2.10 \text{ x } 10^3$ g</u>.

10. To solve this problem, we must first find the balanced chemical equation for this reaction. According to the problem, the reaction is the decomposition of dinitrogen pentaoxide:

$$N_2O_5 \rightarrow N_2 + O_2$$

Balancing the equation gives us:

$$2N_2O_5 \rightarrow 2N_2 + 5O_2$$

The chemical equation, then, tells us:

$$2 \text{ moles } N_2O_5 = 5 \text{ moles } O_2$$

Now all we have to do is use this fact in a conversion:

$$\frac{1.2 \cancel{\text{moles} N_2O_5}}{1} \times \frac{5 \text{ moles } O_2}{2 \cancel{\text{moles} N_2O_5}} = 3.0 \text{ moles} O_2$$

Thus, <u>3.0 moles of oxygen</u> will be formed .

SAMPLE CALCULATIONS FOR EXPERIMENT 5.1

Number of drops to get 10.0 mL: 251

mL per drop: $\frac{10.0 \text{ mL}}{251 \text{ drops}} = 0.0398 \frac{\text{mL}}{\text{drop}}$

Diameter of the circle: 12.75 cm

The volume of one drop is 0.0398 mL. To get the mass in one drop, we just multiply that by the density of the solution, which was given:

$$0.0398 \cancel{\text{mL}} \times 1.00 \frac{\text{g}}{\cancel{\text{mL}}} = 0.0398 \text{ g}$$

That's the mass of the solution, but only a fraction of that is sodium stearate:

$$\text{Mass of sodium sterate added} = 0.0398 \text{ g} \times 0.000125 = 4.98 \times 10^{-6} \text{ g}$$

Now we need to convert to moles. First, then, we need to know the mass of a sodium stearate molecule:

$$1 \text{ x } 23.0 \text{ amu} + 18 \text{ x } 12.0 \text{ amu} + 35 \text{ x } 1.01 \text{ amu} + 2 \text{ x } 16.0 \text{ amu} = 306.4 \text{ amu}$$

This means:

$$1 \text{ mole } NaC_{18}H_{35}O_2 = 306.4 \text{ g } NaC_{18}H_{35}O_2$$

We can use that to convert to moles:

$$\frac{4.98 \times 10^{-6} \cancel{\text{g } NaC_{18}H_{35}O_2}}{1} \times \frac{1 \text{ mole } NaC_{18}H_{35}O_2}{306.4 \cancel{\text{g } NaC_{18}H_{35}O_2}} = 1.63 \times 10^{-8} \text{ moles } NaC_{18}H_{35}O_2$$

Since a mole contains $6.02 \text{ x } 10^{23}$ molecules, we can determine the number of molecules:

$$\text{Number of molecules} = (1.63 \times 10^{-8} \cancel{\text{moles}}) \times (6.02 \times 10^{23} \frac{\text{molecules}}{\cancel{\text{mole}}}) = 9.81 \times 10^{15} \text{ molecules}$$

Now we need to know the area of the circle:

$$A = \pi \cdot r^2 = (3.1416) \text{ x } (6.375 \text{ cm})^2 = 127.7 \text{ cm}^2$$

Now we can calculate the area per molecule:

$$\frac{127.7 \text{ cm}^2}{9.81 \times 10^{15}} = 1.30 \times 10^{-14} \text{ cm}^2$$

The width is roughly the square root of that, or <u>$1.14 \text{ x } 10^{-7}$ cm</u>.

SOLUTIONS TO THE PRACTICE PROBLEMS FOR MODULE #6

1. To solve this problem, we must first come up with the chemical equation. This can be done because our definition of complete combustion gives us the unbalanced equation:

$$C_3H_8 + O_2 \rightarrow CO_2 + H_2O$$

Balancing:

$$C_3H_8 + 5O_2 \rightarrow 3CO_2 + 4H_2O$$

Now we can do stoichiometry. Since C_3H_8 is the limiting reactant, it determines the amount of product formed:

$$1 \text{ mole } C_3H_8 = 3 \text{ moles } CO_2$$

$$\frac{2.13 \cancel{\text{moles } C_3H_8}}{1} \times \frac{3 \text{ moles } CO_2}{1 \cancel{\text{mole } C_3H_8}} = 6.39 \text{ moles } CO_2$$

There will be 6.39 moles of carbon dioxide formed.

2. This problem asks us to relate the quantity of the product back to the limiting reactant. According to the equation:

$$2 \text{ moles } H_2S = 2 \text{ moles } Ag_2S$$

Thus

$$\frac{0.0012 \cancel{\text{moles } Ag_2S}}{1} \times \frac{2 \text{ moles } H_2S}{2 \cancel{\text{moles } Ag_2S}} = 0.0012 \text{ moles } H_2S$$

This means the silver was exposed to 0.0012 moles of H_2S.

3. This problem also relates the quantity of a product back to the amount of reactant.

$$3 \text{ moles } CCl_2F_2 = 2 \text{ moles } SbF_3$$

$$\frac{1.0 \times 10^4 \cancel{\text{moles } CCl_2F_2}}{1} \times \frac{2 \text{ moles } SbF_3}{3 \cancel{\text{moles } CCl_2F_2}} = 6.7 \times 10^3 \text{ moles } SbF_3$$

The chemist needs 6.7×10^3 moles of antimony trifluoride.

4. Since we are using liters and the two substances in the problem are gases, we can use Gay-Lussac's Law and relate the liters of each substance:

$$3 \text{ liters } CCl_2F_2 = 3 \text{ liters } CCl_4$$

$$\frac{100.0 \cancel{\text{L } CCl_2F_2}}{1} \times \frac{3 \text{ L } CCl_4}{3 \cancel{\text{L } CCl_2F_2}} = 100.0 \text{ L } CCl_4$$

The chemist needs 100.0 liters of carbon tetrachloride.

5. This problem asks us to relate the quantity of limiting reactant to the quantity of product, but the quantities are both in grams. So first we must convert to moles:

$$\frac{1.50 \times 10^4 \ \cancel{g\,H_2SO_4}}{1} \times \frac{1 \text{ mole } H_2SO_4}{98.1 \ \cancel{g\,H_2SO_4}} = 153 \text{ moles } H_2SO_4$$

Now we can use stoichiometry to relate the two substances :

$$\frac{153 \ \cancel{\text{moles } H_2SO_4}}{1} \times \frac{1 \text{ moles} CaH_4P_2O_8}{2 \ \cancel{\text{moles} H_2SO_4}} = 76.5 \text{ moles} CaH_4P_2O_8$$

Now that we know how much fertilizer was made, we just need to get it in grams:

$$\frac{76.5 \ \cancel{\text{moles} CaH_4P_2O_8}}{1} \times \frac{234.1 \text{ g} CaH_4P_2O_8}{1 \ \cancel{\text{mole} CaH_4P_2O_8}} = 1.79 \times 10^4 \text{ g} CaH_4P_2O_8$$

Thus, $\underline{1.79 \text{ x } 10^4 \text{ g of fertilizer}}$ will be made.

6. We assume that 10.0 g of HCl is deadly. If we relate that back to Phosgene through the chemical equation, we can determine how many grams of Phosgene is deadly. The problem is, to do that, we must first convert to moles, because stoichiometry can only be done in moles:

$$\frac{10.0 \ \cancel{g\ HCl}}{1} \times \frac{1 \text{ mole HCl}}{36.5 \ \cancel{g\ HCl}} = 0.274 \text{ moles HCl}$$

Now we can convert from HCl to Phosgene:

$$\frac{0.274 \ \cancel{\text{moles HCl}}}{1} \times \frac{1 \text{ mole } COCl_2}{2 \ \cancel{\text{moles HCl}}} = 0.137 \text{ moles} COCl_2$$

Now we know how much Phosgene. All we need to do now is convert to grams:

$$\frac{0.137 \ \cancel{\text{moles} COCl_2}}{1} \times \frac{99.0 \text{ g} COCl_2}{1 \ \cancel{\text{mole} COCl_2}} = 13.6 \text{ g} COCl_2$$

This means that only $\underline{13.6 \text{ g of } COCl_2}$ gas is deadly.

7. This problem asks us to relate the amount of one reactant to the amount of another. This is easy, as long as we start with moles, not grams:

$$\frac{3.6 \times 10^5 \ \cancel{g\ Mg}}{1} \times \frac{1 \text{ mole Mg}}{24.3 \ \cancel{g\ Mg}} = 1.5 \times 10^4 \text{ moles Mg}$$

Now we can convert from Mg to water:

$$\frac{1.5 \times 10^4 \text{ \sout{moles Mg}}}{1} \times \frac{2 \text{ moles } H_2O}{1 \text{ \sout{mole Mg}}} = 3.0 \times 10^4 \text{ moles } H_2O$$

Now we know how much water is needed. All we need to do now is convert to grams:

$$\frac{3.0 \times 10^4 \text{ \sout{moles } \sout{H_2O}}}{1} \times \frac{18.0 \text{ g } H_2O}{1 \text{ \sout{mole } \sout{H_2O}}} = 5.4 \times 10^5 \text{ g } H_2O$$

Thus, 5.4×10^5 g of water must be used.

8. The mass of the empirical formula is:

Mass $CHBr_2$ = 1 x 12.0 amu + 1 x 1.01 amu + 2 x 79.9 amu = 172.8 amu

In order to get that equal to the molecular mass, we must multiply it by 2. Thus, the empirical formula must be multiplied by 2 as well:

$$C_{1x2}H_{1x2}Br_{2x2} = C_2H_2Br_4$$

9. The unbalanced equation for the decomposition is:

$$C_xH_yO_z \rightarrow C + H_2 + O_2$$

To get the stoichiometric coefficients on the products side, we use the experimental data:

$$\frac{63.2 \text{ \sout{g C}}}{1} \times \frac{1 \text{ mole C}}{12.0 \text{ \sout{g C}}} = 5.27 \text{ moles C}$$

$$\frac{5.26 \text{ \sout{g } \sout{H_2}}}{1} \times \frac{1 \text{ mole } H_2}{2.02 \text{ \sout{g } \sout{H_2}}} = 2.60 \text{ moles } H_2$$

$$\frac{41.6 \text{ \sout{g } \sout{O_2}}}{1} \times \frac{1 \text{ mole } O_2}{32.0 \text{ \sout{g } \sout{O_2}}} = 1.30 \text{ moles } O_2$$

So the equation becomes:

$$C_xH_yO_z \rightarrow 5.27C + 2.60H_2 + 1.30O_2$$

We now must divide by the smallest number to make them integers:

$$C_xH_yO_z \rightarrow \frac{5.27}{1.30}C + \frac{2.60}{1.30}H_2 + \frac{1.30}{1.30}O_2$$

$$C_xH_yO_z \rightarrow 4C + 2H_2 + O_2$$

To balance the equation, then, x=4, y=4, z=2. This makes a formula of $C_4H_4O_2$, but that is not an empirical formula because the subscripts have a common factor of 2. Thus, the real empirical formula is C_2H_2O.

10. This is actually two problems rolled into one. First, you must find the empirical formula, then you must use the molecular mass to find the molecular formula:

$$C_xCl_y \rightarrow C + Cl_2$$

$$\frac{14.5\ \cancel{g C}}{1} \times \frac{1\ \text{mole C}}{12.0\ \cancel{g C}} = 1.21\ \text{moles C}$$

$$\frac{85.5\ \cancel{g Cl_2}}{1} \times \frac{1\ \text{mole } Cl_2}{71.0\ \cancel{g Cl_2}} = 1.20\ \text{moles } Cl_2$$

$$C_xCl_y \rightarrow 1.21C + 1.20Cl_2$$

Dividing both number by the smallest yields:

$$C_xCl_y \rightarrow C + Cl_2$$

For the equation to balance, the formula is:

$$CCl_2$$

To determine the molecular formula:

$$\text{Mass of } CCl_2 = 1 \text{ x } 12.0 \text{ amu} + 2 \text{ x } 35.5 \text{ amu} = 83.0 \text{ amu}$$

To get that to equal the molar mass, we must multiply by 2. Therefore, we also must multiply the empirical formula by 2:

$$\underline{C_2Cl_4}$$

SOLUTIONS TO THE PRACTICE PROBLEMS FOR MODULE #7

1. a. Looking at the chart, Zr has an atomic number of 40. This means it has 40 protons and 40 electrons. Its mass number, according to the problem, is 90. If it has 90 total protons + neutrons, and it has 40 protons, then it has 90 - 40 = 50 neutrons.

   b. Looking at the chart, mercury (Hg) has an atomic number of 80. This means it has 80 protons and 80 electrons. Its mass number, according to the problem, is 202. If it has 202 total protons + neutrons, and it has 80 protons, then it has 202 - 80 = 122 neutrons.

   c. Looking at the chart, Ni has an atomic number of 28. This means it has 28 protons and 28 electrons. Its mass number, according to the problem, is 58. If it has 58 total protons + neutrons, and it has 28 protons, then it has 58 - 28 = 30 neutrons.

   d. Looking at the chart, Rn has an atomic number of 86. This means it has 86 protons and 86 electrons. Its mass number, according to the problem, is 222. If it has 222 total protons + neutrons and it has 86 protons, then it has 222 - 86 = 136 neutrons.

2. Isotopes have the name number of protons (thus the same atomic number and the same element symbol) but different numbers of neutrons (thus different mass numbers). Therefore ^{22}Na, ^{23}Na, and ^{24}Na are isotopes. They all have the same number of protons (11), but they have 11, 12, and 13 neutrons, respectively. Remember, isotope is a relational term. It tells you how atoms relate to one another. There is not one "normal" atom with the rest being isotopes. Any group of atoms that all have the same number of protons but different numbers of neutrons are isotopes.

3. If it has 39 protons and electrons, its atomic number is 39. The symbol that has atomic number 39 is Y. The mass number is the number of protons plus the number of neutrons, or 39 + 45 = 84. Thus, the symbol is ^{84}Y.

4. To solve this, we use Equation (7.1):

$$f = \frac{c}{\lambda}$$

But now we have to rearrange it so that we are solving for wavelength:

$$\lambda = \frac{c}{f}$$

Now we can plug in the numbers:

$$\lambda = \frac{3.0 \times 10^{8}\ \frac{m}{s}}{1.2 \times 10^{14}\ \frac{1}{s}} = 2.5 \times 10^{-6}\ m$$

The wavelength is 2.5×10^{-6} m. Each number in the equation has two significant figures, so the answer must have two significant figures.

5. This problem is a direct application of Equation (7.2):

$$E = h \cdot f$$

$$E = 6.63 \times 10^{-34} \frac{J}{\cancel{Hz}} \cdot 5.3 \times 10^{20}\ \cancel{Hz} = 3.5 \times 10^{-13}\ J$$

The energy is $\underline{3.5 \times 10^{-13}\text{ Joules}}$.

6. This problem is a little difficult because the only equation we can use to calculate the energy of light is Equation (7.2), and it uses *frequency*, not *wavelength*. In order to solve this, then, we must first turn the wavelength we've been given into frequency. We can do this with Equation (7.1). We first, however, must get units to work out. Since Equation (7.1) uses c, and c is in m/s, we need to get our wavelength in meters before we can use Equation (7.1).

$$\frac{351\ \cancel{nm}}{1} \times \frac{10^{-9}\ m}{1\ \cancel{nm}} = 3.51 \times 10^{-7}\ m$$

Now we can use Equation (7.1):

$$f = \frac{c}{\lambda}$$

$$f = \frac{3.0 \times 10^{8}\ \frac{\cancel{m}}{s}}{3.51 \times 10^{-7}\ \cancel{m}} = 8.5 \times 10^{14}\ \frac{1}{s} = 8.5 \times 10^{14}\ Hz$$

Now that we have the frequency, we can use Equation (7.2):

$$E = h \cdot f$$

$$E = 6.63 \times 10^{-34} \frac{J}{\cancel{Hz}} \cdot 8.5 \times 10^{14}\ \cancel{Hz} = 5.6 \times 10^{-19}\ J$$

The energy is $\underline{5.6 \times 10^{-19}\text{ Joules}}$.

7. The electron in the 3s orbital has the higher energy, because energy level 3 is higher in energy than energy level 2. The electron in the 2p orbital is orbiting the nucleus in a dumbbell shape, while the 3s electron is orbiting the nucleus in a spherical shape.

8. a. To get to element Ti, we must go through row 1, which has two boxes in the s orbital block ($1s^2$). We then go through all of row 2, which has 2 boxes in the s orbital block and 6 boxes in the p orbital block ($2s^22p^6$). We also go through row 3, which has two boxes in the s orbital block and 6 in the p orbital block ($3s^23p^6$). We then go to the fourth row, where we pass through both boxes in the s orbital

block ($4s^2$). Finally, we go through 2 boxes in the d orbital block. Since we subtract one from the row number for d orbitals, this gives us $3d^2$. Thus, our final electron configuration is:

$$\underline{1s^2 2s^2 2p^6 3s^2 3p^6 4s^2 3d^2}$$

b. To get to element S, we must go through row 1, which has two boxes in the s orbital block ($1s^2$). We then go through all of row 2, which has 2 boxes in the s orbital block and 6 boxes in the p orbital block ($2s^2 2p^6$). We also go through both boxes in the s orbital block of row 3, ($3s^2$). Finally, we go through 4 boxes in the p orbital block of row 3, giving us $3p^4$. Thus, our final electron configuration is:

$$\underline{1s^2 2s^2 2p^6 3s^2 3p^4}$$

c. To get to element Rb, we must go through row 1, which has two boxes in the s orbital block ($1s^2$). We then go through all of row 2, which has 2 boxes in the s orbital block and 6 boxes in the p orbital block ($2s^2 2p^6$). We also go through row 3, which has two boxes in the s orbital block and 6 in the p orbital block ($3s^2 3p^6$). We then go to the fourth row, where we pass through both boxes in the s orbital block, all 10 boxes in the d orbital block, and all 6 boxes in the p orbital block. Since we subtract one from the row number for d orbitals, this gives us $4s^2 3d^{10} 4p^6$. Finally, we end up in the first box of the row 5, s orbital block. Thus, our final electron configuration is:

$$\underline{1s^2 2s^2 2p^6 3s^2 3p^6 4s^2 3d^{10} 4p^6 5s^1}$$

9. a. The nearest 8A element that has a lower atomic number than V is Ar. The only difference between V and Ar is that there are 2 boxes in the row 4, s orbital group and 3 boxes in the row 4, d orbital group. Therefore, the abbreviated electron configuration for V is:

$$\underline{[Ar]4s^2 3d^3}$$

b. The nearest 8A element that has a lower atomic number than Sn is Kr. The only difference between Sn and Kr is that there are 2 boxes in the row 5, s orbital group, 10 boxes in the row 5, d orbital group, and 2 boxes in the row 5, p orbital group. Thus, the abbreviated electron configuration for Sn is:

$$\underline{[Kr]5s^2 4d^{10} 5p^2}$$

c. The nearest 8A element that has a lower atomic number than In is Kr. The only difference between In and Kr is that there are 2 boxes in the row 5, s orbital group, 10 boxes in the row 5, d orbital group, and 1 box in the row 5, p orbital group. Thus, the abbreviated electron configuration for In is:

$$\underline{[Kr]5s^2 4d^{10} 5p^1}$$

10. a. <u>You cannot have 7 electrons in p orbitals. Since there are three p orbitals per energy level and each can contain 2, the most you can ever have is 6</u>.

b. <u>The order that the orbitals were filled is wrong. 3d fills up after 4s</u>.

SOLUTIONS TO THE PRACTICE PROBLEMS FOR MODULE #8

1. a. Ge is in group 4A, so it has 4 valence electrons:

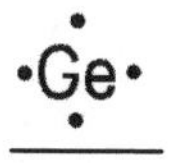

```
  •
•Ge•
  •
```

b. Te is in group 6A, so it has 6 valence electrons:

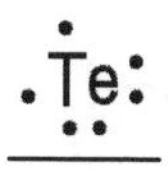

```
  •
.Te:
  ••
```

c. Ba is in group 2A, so it has 2 valence electrons:

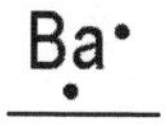

```
Ba•
 •
```

2. a. Al is in group 3A, so it wants a charge of 3+. Sulfur is in group 6A, so sulfide will have a charge of 2-. Ignoring the signs and switching the numbers gives us Al_2S_3.

b. Cs is in group 1A, so it wants a charge of 1+. Nitrogen is in group 5A, so nitride will have a charge of 3-. Ignoring the signs and switching the numbers gives us Cs_3N.

c. Mg is in group 2A, so it wants a charge of 2+. Oxygen is in group 6A, so oxide will have a charge of 2-. The numbers are the same so we ignore them: MgO.

d. Cr is an exception, because there is a Roman numeral in the name. The numeral tells us that Cr want a charge of 3+. Oxygen is in group 6A, so oxide will have a charge of 2-. Ignoring the signs and switching the numbers gives us Cr_2O_3.

3. Ionization potential decreases as you go down the chart. The atom with the lowest ionization potential will give up its electrons easiest. That would be In.

4. Ionization potential increases as you go from left to right on the chart, so the order is Sr < Sb < I.

5. Electronegativity decreases as you go down the chart, so N has the greatest desire for extra electrons.

6. Atomic radius decreases as you go from left to right on the chart, so the order is Br < Se < As < K.

7. The chemical formula tells us that we have one C and four H's to work with:

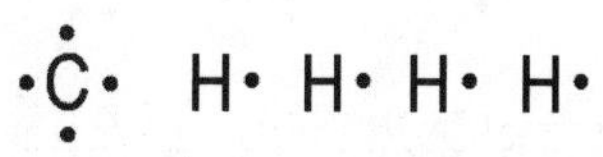

```
 •
•C•   H•  H•  H•  H•
 •
```

Because C has the most unpaired electrons, it goes in the center and we try to attach the H's to it. This is easy since each H has a space for an unpaired electron, and the C has four unpaired electrons. The Lewis structure, then, looks like this:

H
H:C:H
H

All atoms have their ideal electron configuration, so we are all set. Now we just have to replace the shared electron pairs with dashes:

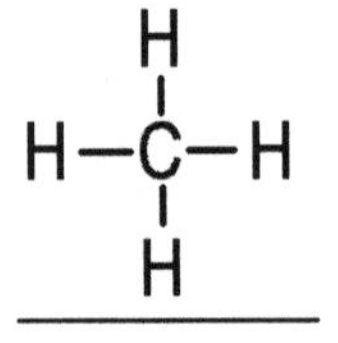

8. The chemical formula tells us that we have one P and three Cl's to work with:

·P: :Cl: :Cl: :Cl:

Because P has the most unpaired electrons, it goes in the center and we try to attach the Cl's to it. This is easy since each Cl has a space for an unpaired electron, and the P has three unpaired electrons. The Lewis structure, then, looks like this:

:Cl:
:Cl:P:
:Cl:

All atoms have their ideal electron configuration, so we are all set. Now we just have to replace the shared electron pairs with dashes:

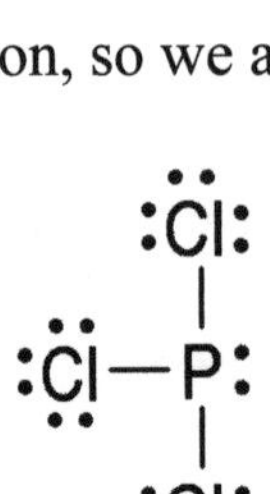

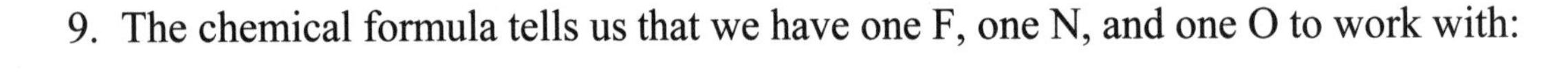

9. The chemical formula tells us that we have one F, one N, and one O to work with:

:F: ·N: ·O:

The N has the most unpaired electrons, so it goes in the middle. We attach the others to it:

:F:N:O:

The F now has eight electrons, so its all set. The N and O, however, have only seven each. We will give the O eight by taking the unpaired electron on the N and putting it in between the N and the O:

$$:\ddot{\underset{..}{F}}:\dot{\underset{..}{N}}:\dot{\underset{..}{O}}: \longrightarrow :\ddot{\underset{..}{F}}:\underset{..}{N}:.\dot{\underset{..}{O}}:$$

We can also give the N its eight by taking the unpaired electron on the oxygen and moving it in between the N and the O.

$$:\ddot{\underset{..}{F}}:N::\dot{\underset{..}{O}}: \longrightarrow :\ddot{\underset{..}{F}}:\underset{..}{N}::\underset{..}{O}:$$

Now all atoms have eight valence electrons. All we have to do is replace the shared electron pairs with dashes:

$$:\ddot{\underset{..}{F}}-\underset{..}{N}=\underset{..}{O}:$$

10. To solve this, we have to determine the Lewis structures of each molecule. They turn out to be:

$$H-\underset{\substack{|\\H}}{C}=\underset{\substack{|\\H}}{C}-H \qquad :P\equiv N: \qquad H-H$$

Based on these three Lewis structures, the <u>H_2 will be easiest to break apart</u> because it is held together by a single bond, while the C_2H_4 is held together by a double bond, and the PN is held together by a triple bond.

SOLUTIONS TO THE PRACTICE PROBLEMS FOR MODULE #9

1. Ionic compounds are named by simply listing the ions present. In order to get the formula, you must determine the charge of each ion and balance those charges. We learned how to do this in the last module, so the only new thing here is the fact that there are now polyatomic ions to consider.

a. The name indicates a potassium ion and a sulfate ion. Potassium is abbreviated with a K, and, since it is in group 1A, it has a charge of 1+. We are supposed to have memorized that the sulfate ion is SO_4 and has a charge of 2-. Ignoring the signs and switching the numbers gives us:

$$\underline{K_2SO_4}$$

We do not put parentheses around the polyatomic ion because there is only one sulfate ion in the molecule.

b. The name indicates a calcium ion and a nitrate ion. Calcium is abbreviated with a Ca, and, since it is in group 2A, it has a charge of 2+. We are supposed to have memorized that the nitrate ion is NO_3 and has a charge of 1-. Ignoring the signs and switching the numbers gives us:

$$\underline{Ca(NO_3)_2}$$

c. The name indicates a magnesium ion and a carbonate ion. Magnesium is abbreviated with a Mg, and, since it is in group 2A, it has a charge of 2+. We are supposed to have memorized that the carbonate ion is CO_3 and has a charge of 2-. Since the numerical values of the charges are the same, we ignore them. This gives us:

$$\underline{MgCO_3}$$

We do not put parentheses around the polyatomic ion because there is only one carbonate ion in the molecule.

d. The name indicates an aluminum ion and a chromate ion. Aluminum is abbreviated with an Al, and, since it is in group 3A, it has a charge of 3+. We are supposed to have memorized that the chromate ion is CrO_4 and has a charge of 2-. Ignoring the signs and switching the numbers gives us:

$$\underline{Al_2(CrO_4)_3}$$

2. In order to name ionic compounds, we only have to put the names of the ions together.

a. Since we see that NH_4 is in parentheses, that means it is a polyatomic ion. We are supposed to have memorized that NH_4^+ is the ammonium ion, and the only other ion is the single-atom oxide ion. Thus, the name is ammonium oxide.

b. In looking at this molecule, we should notice the NO_2. It tells us the nitrite polyatomic ion is present. The only thing left after that is the potassium ion. Thus, the name is potassium nitrite.

c. Since we see that PO_4 is in parentheses, that means it is a polyatomic ion. We are supposed to have memorized that PO_4^{3-} is the phosphate ion, and the only other ion is the single-atom calcium ion. Thus, the name is calcium phosphate.

d. In looking at this molecule, we should notice the PO_4. It tells us that the phosphate polyatomic ion is present. The only thing left after that is the aluminum ion. Thus, the name is aluminum phosphate.

3. First, we have to determine the formulas of the molecules involved:

Calcium nitrate includes Ca^{2+} and NO_3^-. Thus, its formula is $Ca(NO_3)_2$.
Sodium carbonate includes Na^+ and CO_3^{2-}. Thus, its formula is Na_2CO_3.
Calcium carbonate includes Ca^{2+} and CO_3^{2-}. Thus, it formula is $CaCO_3$.
Sodium nitrate includes $Na^+$and NO_3^-. Thus, its formula is $NaNO_3$.

The problem tells us the phases, so the unbalanced equation is:

$$Ca(NO_3)_2\ (aq) + Na_2CO_3\ (aq) \rightarrow CaCO_3\ (s) + NaNO_3\ (aq)$$

Remembering that you must distribute any superscripts after parentheses to all atoms in the parentheses, the atomic inventory is:

Reactants Side	Products Side
Ca: 1x1 = 1	Ca: 1x1 = 1
N: 1x1x2 = 2	N: 1x1 = 1
Na: 1x2 = 2	Na: 1x1 = 1
C: 1x1 = 1	C: 1x1 = 1
O: 1x3x2 + 1x3 = 9	O: 1x3 + 1x3 = 6

To balance, we need to multiply the $NaNO_3$ by two:

$$\underline{Ca(NO_3)_2\ (aq) + Na_2CO_3\ (aq) \rightarrow CaCO_3\ (s) + 2NaNO_3\ (aq)}$$

If you count it all up now, the equation balances.

4. To determine shapes, we must first draw the Lewis structure:

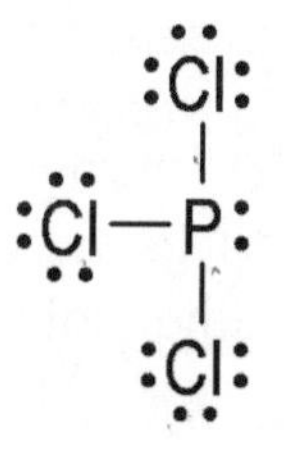

We see that the central atom has four groups of electrons around it. Three of them are bonds, and one is a non-bonding pair. Since there are four groups, the basic shape is that of a tetrahedron. However, one of the legs is missing because it contains a non-bonding pair of electrons. As a result, the molecule's shape is pyramidal with a bond angle of 107°:

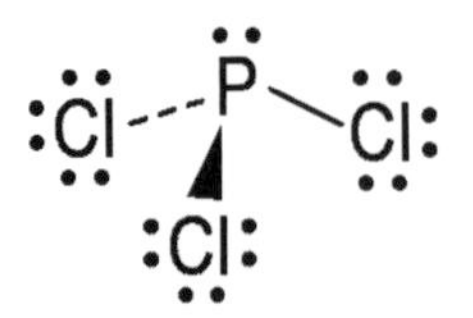

5. To determine shapes, we must first draw the Lewis structure:

H–H

Since there are only two atoms here, the molecule is linear with a bond angle of 180°. The picture looks just like the Lewis structure.

6. To determine shapes, we must first draw the Lewis structure:

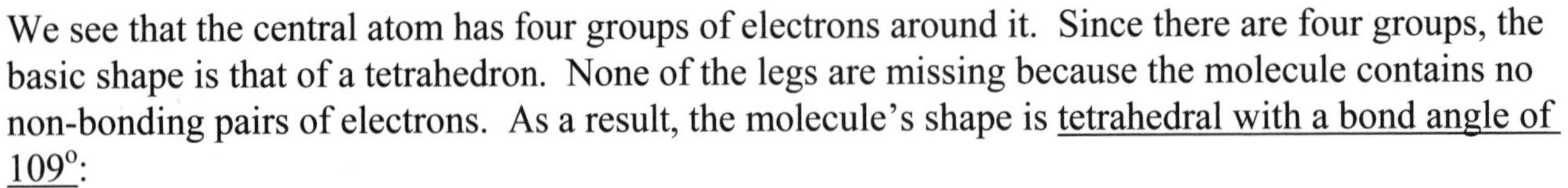

We see that the central atom has four groups of electrons around it. Since there are four groups, the basic shape is that of a tetrahedron. None of the legs are missing because the molecule contains no non-bonding pairs of electrons. As a result, the molecule's shape is tetrahedral with a bond angle of 109°:

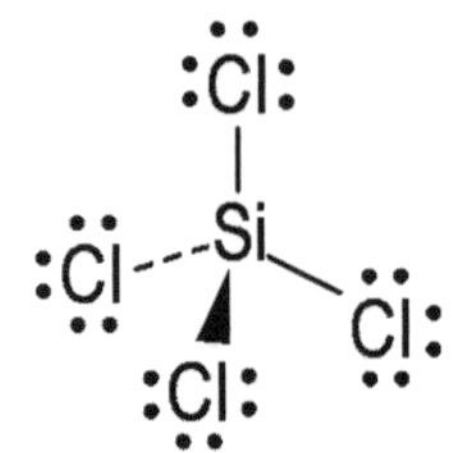

7. To determine shapes, we must first draw the Lewis structure:

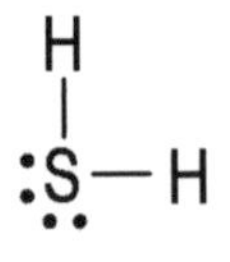

We see that the central atom has four groups of electrons around it. Since there are four groups, the basic shape is that of a tetrahedron. Two of the legs are missing, however, because two of the groups are non-bonding pairs of electrons. As a result, the molecule's shape is bent with a bond angle of 105°:

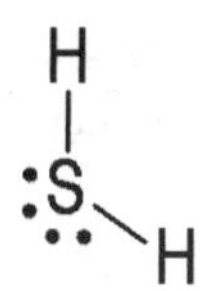

8. To determine shapes, we must first draw the Lewis structure:

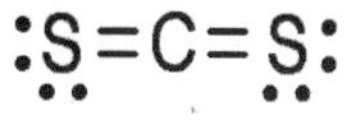

:S=C=S:

We see that the central atom has two groups of electrons around it. Since there are two groups, the shape is linear with a bond angle of 180°. The picture looks just like the Lewis structure.

9. a. This compound contains a metal (Mg) and a nonmetal (Cl). It is therefore an ionic compound.

b. This molecule contains no metals, so it is either purely covalent or polar covalent. To determine which, we must first start with a Lewis structure:

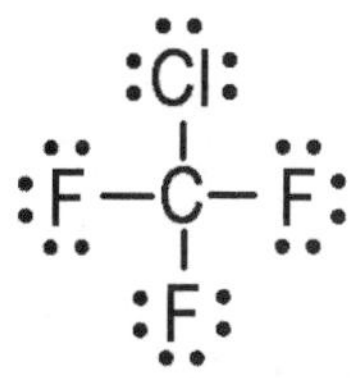

:Cl:
:F—C—F:
:F:

We then determine its shape. Since the central atom is surrounded by four groups of electrons and there are no non-bonding electron pairs on the central atom, the shape is tetrahedral. Now that we know the shape, we can look at the direction that the electrons are being pulled. In this case, both F and Cl are more electronegative than C, so the electrons are all pulled away from the carbon:

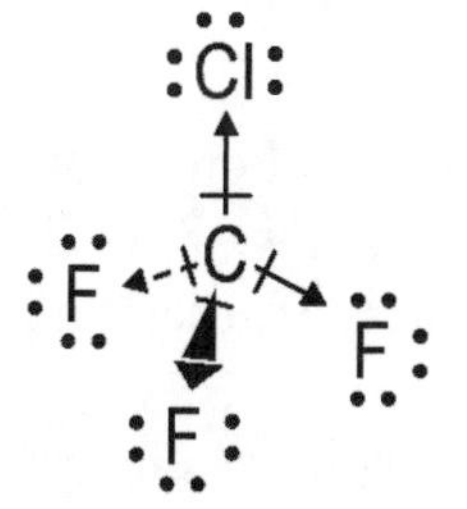

The electrons are all being pulled in opposite directions, but the pulls are not equal. Since F is more electronegative than Cl, it can pull on the electrons harder. Thus, the molecule is polar covalent.

c. We determined the shape of this in a previous problem. We now simply have to determine the direction in which the electrons are being pulled. S is two steps to the right of C on the periodic table and only 1 step down. Thus, even though it loses a little electronegativity by being lower than C on the chart, it gains more because it is two steps to the right of C. As a result, the electrons are pulled towards the sulfurs:

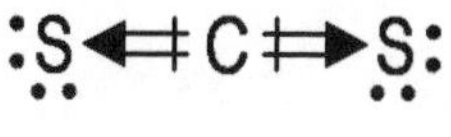

:S⟵C⟶S:

We see from the picture that the electrons are being pulled equally in opposite directions. Thus, the polar bonds cancel each other out and we are left with a purely covalent compound.

d. This is an easy one. There are no polar bonds in this molecule, since the only atoms in it are identical. Since there are no polar bonds, the molecule must be purely covalent.

e. We determined the shape of this in a previous problem. We now simply have to determine the direction in which the electrons are being pulled. Cl is to the right of Si on the periodic table. Thus, the electrons are pulled towards the chlorines:

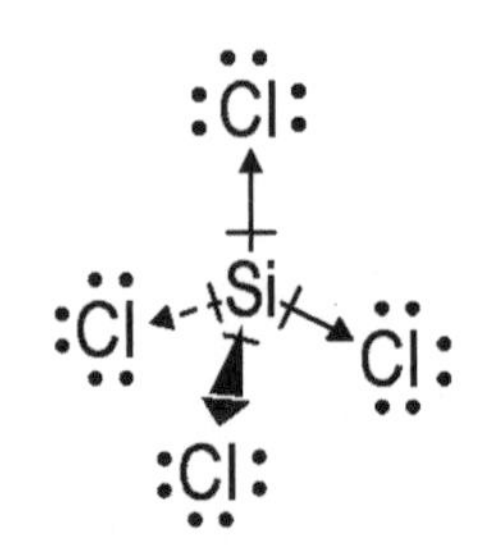

We see from the picture that the electrons are being pulled equally in opposite directions. As a result, the polar bonds cancel each other out, and the molecule is purely covalent.

f. We determined the shape of this in a previous problem. We now simply have to determine the direction in which the electrons are being pulled. Cl is to the right of P on the periodic table. Thus, the electrons are pulled towards the chlorines:

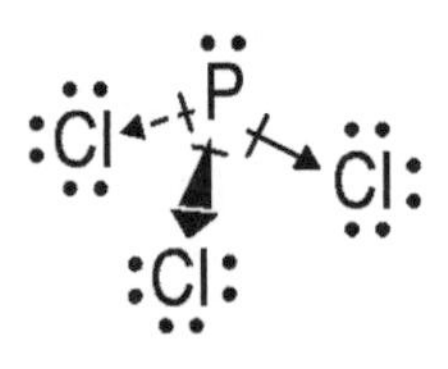

We see from the picture that the electrons are not being pulled in opposite directions, since there is no chlorine pulling straight up. There is simply a pair of non-bonding electrons there. As a result, the polar bonds do not cancel each other out, and the molecule is polar covalent.

10. Water is a polar covalent compound; thus, it has electrical charges. Only other polar covalent compounds or ionic compounds can dissolve in water, because they also have electrical charges in them. Purely covalent compounds cannot dissolve in water. We classified all of the compounds in the previous problem, therefore:

$MgCl_2$, CF_3Cl, and PCl_3 will dissolve in water, the other three will not.

SOLUTIONS TO THE PRACTICE PROBLEMS FOR MODULE #10

1. You should still have Table 9.1 memorized. Thus, you should know that the hydroxide ion is OH^-. Since Al is in group 3A, it takes on a 3+ charge in ionic compounds. Switching the charges and dropping the signs as you learned in Module #9 tells us that aluminum hydroxide is $Al(OH)_3$. Acids and bases usually react to give a salt and water. The salt is made up of the positive ion from the base (Al^{3+}) and the negative ion left over when the acid gets rid of its H^+ ions. In this case, that will be SO_4^{2-}. Switching the charges and dropping the signs gives us $Al_2(SO_4)_3$ as the chemical formula. The unbalanced equation, then, is:

$$H_2SO_4 + Al(OH)_3 \rightarrow H_2O + Al_2(SO_4)_3$$

Now all we have to do is balance it:

$$\underline{3H_2SO_4 + 2Al(OH)_3 \rightarrow 6H_2O + Al_2(SO_4)_3}$$

2. Once again, the hydroxide ion is OH^-. Calcium is in group 2A of the chart and thus takes on a 2+ charge in ionic compounds. This tells us that calcium hydroxide is $Ca(OH)_2$. Acids and bases usually react to give a salt and water. The salt is made up of the positive ion from the base (Ca^{2+}) and the negative ion left over when the acid gets rid of its H^+ ions. In this case, that will be NO_3^-. Switching the charges and dropping the signs tells us that these two ions form $Ca(NO_3)_2$. The unbalanced equation, then, is:

$$HNO_3 + Ca(OH)_2 \rightarrow H_2O + Ca(NO_3)_2$$

Now all we have to do is balance it:

$$\underline{2HNO_3 + Ca(OH)_2 \rightarrow 2H_2O + Ca(NO_3)_2}$$

3. In this case, the base does not contain a hydroxide ion. In fact, it is not even ionic. Thus, a salt and water are not formed in this problem. Here, we just rely on the definition of acids and bases. Ammonia will want to gain an H^+ to become NH_4^+, and the carbonic acid will want to give up both of its H^+ ions. Since the acid has two H^+ ions to give but the base can only accept one, two ammonias will have to do the accepting. This will make two NH_4^+ ions:

$$\underline{H_2CO_3 + 2NH_3 \rightarrow 2NH_4^+ + CO_3^{2-}}$$

4. Water is an amphiprotic substance, so it can act as either an acid or a base. In this problem, it is being mixed with an acid, so it will want to act like a base and accept the H^+ that the acid gives away:

$$\underline{HBr + H_2O \rightarrow H_3O^+ + Br^-}$$

5. a. Molarity is given by number of moles divided by number of liters. We have both those units, so we just divide them:

$$\text{Concentration} = \frac{\text{\# moles}}{\text{\# liters}} = \frac{3.51 \text{ moles } HNO_3}{1.2 \text{ L}} = 2.9 \text{ M}$$

The concentration is $\underline{2.9 \text{ M}}$.

b. In order to get concentration, we must have moles and liters. The problem gives us grams and mL, so we must make two conversions:

$$\text{Molecular Mass of KOH} = 1 \times 39.1\ \text{amu} + 1 \times 16.0\ \text{amu} + 1 \times 1.01\ \text{amu} = 56.1\ \text{amu}$$

$$\text{Therefore, 1 mole KOH} = 56.1\ \text{grams KOH}$$

$$\frac{234.1\ \cancel{\text{g KOH}}}{1} \times \frac{1\ \text{mole KOH}}{56.1\ \cancel{\text{g KOH}}} = 4.17\ \text{moles KOH}$$

$$\frac{345\ \cancel{\text{mL}}}{1} \times \frac{0.001\ \text{L}}{1\ \cancel{\text{mL}}} = 0.345\ \text{L}$$

Now we can calculate molarity:

$$\text{Concentration} = \frac{\#\text{moles}}{\#\text{liters}} = \frac{4.17\ \text{moles KOH}}{0.345\ \text{L}} = 12.1\ \frac{\text{moles KOH}}{\text{L}} = 12.1\ \text{M}$$

The concentration is 12.1 M.

c. In order to get concentration, we must have moles and liters. The problem gives us grams and mL, so we must make two conversions:

$$\text{Molecular Mass of } H_3PO_4 = 3 \times 1.01\ \text{amu} + 1 \times 31.0\ \text{amu} + 4 \times 16.0\ \text{amu} = 98.0\ \text{amu}$$

$$\text{Therefore, 1 mole } H_3PO_4 = 98.0\ \text{grams } H_3PO_4$$

$$\frac{4.1\ \cancel{\text{g } H_3PO_4}}{1} \times \frac{1\ \text{mole } H_3PO_4}{98.0\ \cancel{\text{g } H_3PO_4}} = 0.042\ \text{moles } H_3PO_4$$

$$\frac{45\ \cancel{\text{mL}}}{1} \times \frac{0.001\ \text{L}}{1\ \cancel{\text{mL}}} = 0.045\ \text{L}$$

Now we can calculate molarity:

$$\text{Concentration} = \frac{\#\text{moles}}{\#\text{liters}} = \frac{0.042\ \text{moles } H_3PO_4}{0.045\ \text{L}} = 0.93\ \frac{\text{moles } H_3PO_4}{\text{L}} = 0.93\ \text{M}$$

The concentration is 0.93 M.

6. This is a dilution problem, so we use the dilution equation. M_1 is 12.0 M, we need to determine V_1. M_2 is 3.5 M and $V_2 = 500.0$ mL.

$$M_1V_1 = M_2V_2$$

$$(12.0\ \text{M}) \cdot V_1 = (3.5\ \text{M}) \cdot (500\ \text{mL})$$

$$V_1 = \frac{3.5\ \cancel{\text{M}} \cdot 500\ \text{mL}}{12.0\ \cancel{\text{M}}} = 1.5 \times 10^2\ \text{mL}$$

The chemist must take 1.5×10^2 mL of the original solution and dilute it with enough water to make 500.0 mL of solution.

7. The last part of this problem is a dilution problem, but in order to get the original concentration, we must deal with the first part of the problem, which is a concentration problem.

$$\text{Molecular Mass of NaOH} = 1 \times 23.0\,\text{amu} + 1 \times 16.0\ \text{amu} + 1 \times 1.01\,\text{amu} = 40.0\ \text{amu}$$

$$\text{Therefore, 1 mole NaOH} = 40.0\ \text{grams NaOH}$$

$$\frac{950.0\ \cancel{\text{g NaOH}}}{1} \times \frac{1\ \text{mole NaOH}}{40.0\ \cancel{\text{g NaOH}}} = 23.8\ \text{moles NaOH}$$

$$\text{Concentration} = \frac{\#\text{moles}}{\#\text{liters}} = \frac{23.8\ \text{moles NaOH}}{2.00\ \text{L}} = 11.9\ \frac{\text{moles NaOH}}{\text{L}} = 11.9\ \text{M}$$

Now that we know the concentration of the stock solution, this is just a dilution problem:

$$M_1V_1 = M_2V_2$$

$$(11.9\ \text{M}) \cdot V_1 = (0.10\ \text{M}) \cdot 100.0\ \text{mL}$$

$$V_1 = \frac{0.10\ \cancel{\text{M}} \cdot 100.0\,\text{mL}}{11.9\ \cancel{\text{M}}} = 0.84\ \text{mL}$$

The chemist needs to take 0.84 mL of the stock solution and mix it with enough water to make 100.0 mL of solution.

8. In order to solve any stoichiometry problem, we must first figure out the balanced chemical equation. According to the problem, our reactants are H_2CO_3 and KOH. When they react, a salt and water will be produced. The salt will be composed of the positive ion from the base (K^+) and the negative ion left over when the acid gets rid of all of its H^+ ions. In this case, that will be the CO_3^{2-} ion. Thus, the reaction is:

$$H_2CO_3 + KOH \rightarrow K_2CO_3 + H_2O$$

Of course, this equation is not balanced, so that is the first thing to do:

$$H_2CO_3 + 2KOH \rightarrow K_2CO_3 + 2H_2O$$

Now that we have a balanced equation, we can start the stoichiometry. We were told that there is excess KOH, so we know that H_2CO_3 is the limiting reactant. Therefore, we need to know the moles of carbonic acid in order to be able to predict how much potassium carbonate it produces. To do this, we will take concentration times volume:

$$\frac{4.5\ \text{moles}\ H_2CO_3}{1\ \cancel{L\ H_2CO_3}} \times \frac{0.0500\ \cancel{L\ H_2CO_3}}{1} = 0.23\ \text{moles}\ H_2CO_3$$

We can now use this information to calculate the number of moles of potassium carbonate produced.

$$\frac{0.23\ \cancel{\text{moles}\ H_2CO_3}}{1} \times \frac{1\ \text{mole}\ K_2CO_3}{1\ \cancel{\text{mole}\ H_2CO_3}} = 0.23\ \text{moles}\ K_2CO_3$$

Now that we have the number of moles of potassium carbonate, we can convert back to grams:

$$\frac{0.23\ \cancel{\text{moles}\ K_2CO_3}}{1} \times \frac{138.2\ \text{g}\ K_2CO_3}{1\ \cancel{\text{mole}\ K_2CO_3}} = 32\ \text{g}\ K_2CO_3$$

Therefore, 32 g of potassium carbonate are produced.

9. Remember, titrations are just stoichiometry problems, so first we have to come up with a balanced chemical equation:

$$HCl + NaOH \rightarrow NaCl + H_2O$$

Since the endpoint was reached, we know that there was exactly enough acid added to eat up all of the base. First, then, we calculate how many moles of acid were added:

$$\frac{3.5\ \text{moles}\ HCl}{1\ \cancel{L}} \times \frac{0.0343\ \cancel{L}}{1} = 0.12\ \text{moles}\ HCl$$

We can now use the chemical equation to determine how many moles of base were present:

$$\frac{0.12\ \cancel{\text{moles}\ HCl}}{1} \times \frac{1\ \text{moles}\ NaOH}{1\ \cancel{\text{mole}\ HCl}} = 0.12\ \text{moles}\ NaOH$$

Now that we have the number of moles of base present, we simply divide by the volume of base to get concentration:

$$\text{Concentration} = \frac{\#\ \text{moles}}{\#\ \text{liters}} = \frac{0.12\ \text{moles}\ NaOH}{0.0500\ \text{L}} = 2.4\ \text{M}$$

The NaOH solution has a concentration of 2.4 M.

10. Remember, titrations are just stoichiometry problems, so first we have to come up with a balanced chemical equation:

$$H_2SO_4 + 2KOH \rightarrow K_2SO_4 + 2H_2O$$

Since the endpoint was reached, we know that there was exactly enough base added to eat up all of the acid. First, then, we calculate how many moles of base were added:

$$\frac{1.2 \text{ moles KOH}}{1 \text{ \sout{L}}} \times \frac{0.3451 \text{ \sout{L}}}{1} = 0.41 \text{ moles KOH}$$

We can now use the chemical equation to determine how many moles of acid were present:

$$\frac{0.41 \text{ \sout{moles KOH}}}{1} \times \frac{1 \text{ mole } H_2SO_4}{2 \text{ \sout{mole KOH}}} = 0.21 \text{ moles } H_2SO_4$$

Now that we have the number of moles of acid present, we simply divide by the volume of acid to get concentration:

$$\text{Concentration} = \frac{\text{\# moles}}{\text{\# liters}} = \frac{0.21 \text{ moles } H_2SO_4}{0.5000 \text{ L}} = 0.42 \text{ M}$$

The H_2SO_4 has a concentration of <u>0.42 M</u>.

SAMPLE CALCULATIONS FOR EXPERIMENT 10.2

Mass of Graduated Cylinder: 254 g
Mass of Vinegar + Cylinder: 303 g
Mass of Vinegar: 303 g -254 g = 49 g
Density of Vinegar: $\frac{49 \text{ g}}{50.0 \text{ mL}} = 0.98 \frac{\text{g}}{\text{mL}}$
Volume of Vinegar added in Rough Titration: 22.4 mL
Volume of Vinegar added in Careful Titration: 20.8 mL

We start with the grams of vinegar added :

$$20.8 \cancel{\text{mL}} \times 0.98 \frac{\text{g}}{\cancel{\text{mL}}} = 2.0 \times 10^1 \text{ g}$$

Note that the only way to properly report the significant figures is to use scientific notation. Now we have to determine the mass of acid in that mass of vinegar:

$$\text{Mass of acid used} = (\text{Mass of vinegar}) \text{ x } 0.0500 = (2.0 \text{ x } 10^1 \text{ g})\cdot 0.0500 = 1.0 \text{ g}$$

The chemical formula of the acid ($C_2H_4O_2$) tells us that a single molecule has a mass of 60.0 amu. That also tells us that 60.0 g $C_2H_4O_2$ = 1 mole $C_2H_4O_2$.

$$\frac{1.0 \cancel{\text{g}\, C_2H_4O_2}}{1} \times \frac{1 \text{ mole } C_2H_4O_2}{60.0 \cancel{\text{g}\, C_2H_4O_2}} = 0.017 \text{ moles } C_2H_4O_2$$

Now we can determine the amount of base:

$$\frac{0.017 \cancel{\text{moles } C_2H_4O_2}}{1} \times \frac{1 \text{ mole } NH_3}{1 \cancel{\text{mole } C_2H_4O_2}} = 0.017 \text{ moles } NH_3$$

We used 10.0 mL of ammonia, so the concentration of the ammonia is:

$$\text{Concentration} = \frac{\#\text{ moles}}{\#\text{ liters}} = \frac{0.017 \text{ moles } NH_3}{0.010 \text{ L}} = 1.7 \text{ M}$$

PLEASE NOTE: Your answer may be quite different from mine, as the ammonia solution you used might have been stronger or more dilute than the one I used.

SOLUTIONS TO THE PRACTICE PROBLEMS FOR MODULE #11

1. If the hot tub cools down, the solubility of the solutes will decrease. If it gets cold enough, the solubilities might get so low that some of the solute must exit the solution, forming a precipitate.

2. A soda pop gets flat because the CO_2 dissolved in it escapes. Since the solubility of gases increase with decreasing temperature, a colder soda pop will hold onto more CO_2 than a warm soda pop.

3. This is just a stoichiometry problem. We can tell this by the fact that we are being asked to determine the amount of one substance when we are given the amount of another substance. The only way to do that is by stoichiometry. Now, in order to do stoichiometry, we must first get our amount in moles. Right now, the amount is given in concentration and volume. We therefore must turn that into moles. In order to do that, though, we must convert mL into liters, so that our volumes units are consistent with our concentration unit (remember, "M" means $\frac{\text{moles}}{\text{liter}}$).

$$\frac{345\ \cancel{\text{mL}}}{1} \times \frac{0.001\text{ L}}{1\ \cancel{\text{mL}}} = 0.345\text{ L}$$

Now we can convert from volume and concentration into moles:

$$\frac{1.25\text{ moles Al(NO}_3)_3}{1\ \cancel{\text{L}}} \times 0.345\ \cancel{\text{L}} = 0.431\text{ moles Al(NO}_3)_3$$

Now that we have moles, we can do stoichiometry:

$$\frac{0.431\ \cancel{\text{moles Al(NO}_3)_3}}{1} \times \frac{1\text{ mole Al}_2(\text{SO}_4)_3}{2\ \cancel{\text{moles Al(NO}_3)_3}} = 0.216\text{ moles Al}_2(\text{SO}_4)_3$$

Now, of course, this is not quite the answer we need. We were asked to figure out how many grams of aluminum sulfate were produced, so we have to convert from moles back to grams:

$$\frac{0.216\ \cancel{\text{moles Al}_2(\text{SO}_4)_3}}{1} \times \frac{342.3\text{ grams Al}_2(\text{SO}_4)_3}{1\ \cancel{\text{mole Al}_2(\text{SO}_4)_3}} = 73.9\text{ grams Al}_2(\text{SO}_4)_3$$

This means that 73.9 g of aluminum sulfate will be produced.

4. Once again, we can tell that this is a stoichiometry problem because we are asked to convert from amount of one substance to amount of another. In order to do stoichiometry, however, we must convert to moles. Since we have the number of grams of silver bromide, we will use that as our starting point:

$$\frac{100.0\ \cancel{\text{g AgBr}}}{1} \times \frac{1\text{ mole AgBr}}{187.8\ \cancel{\text{g AgBr}}} = 0.5325\text{ moles AgBr}$$

Now that we have moles, we can use stoichiometry to determine how much $AgNO_3$ is needed:

$$\frac{0.5325\ \cancel{\text{moles AgBr}}}{1} \times \frac{1\ \text{moles AgNO}_3}{1\ \cancel{\text{mole AgBr}}} = 0.5325\ \text{moles AgNO}_3$$

This tells us how much $AgNO_3$, but it doesn't answer the question. The question asks how many liters of a 2.5 M solution is needed. To determine this, we must remember that "M" means moles per liter. Thus, the concentration of a solution is a conversion relationship that allows us to relate the number of moles to the number of liters. We can therefore do the following conversion:

$$\frac{0.5325\ \cancel{\text{moles AgNO}_3}}{1} \times \frac{1\ \text{L AgNO}_3}{2.5\ \cancel{\text{moles AgNO}_3}} = 0.21\ \text{L of solution}$$

Thus, 0.21 liters of silver nitrate must be added.

5. In this problem, we can use the number of grams of $PbCl_2$ to determine the amount of Pb^{2+} that was there. First, however, we need to get to the number of moles of lead chloride:

$$\frac{25.0\ \cancel{\text{g PbCl}_2}}{1} \times \frac{1\ \text{mole PbCl}_2}{278.2\ \cancel{\text{g PbCl}_2}} = 0.0899\ \text{moles PbCl}_2$$

Now we can use stoichiometry to determine the number of moles of Pb^{2+} that were in the water sample:

$$\frac{0.0899\ \cancel{\text{moles PbCl}_2}}{1} \times \frac{1\ \text{mole Pb}^{2+}}{1\ \cancel{\text{mole PbCl}_2}} = 0.0899\ \text{moles Pb}^{2+}$$

Now that we have the number of moles of Pb^{2+}, we can divide by the volume of the solution to get molarity:

$$\text{Molarity} = \frac{\#\ \text{moles}}{\#\ \text{liters}} = \frac{0.0899\ \text{moles}}{1.6\ \text{L}} = 0.056\ \text{M}$$

The concentration of Pb^{2+} in the water is 0.056 M.

6. $$\text{molality} = \frac{\#\ \text{moles solute}}{\#\ \text{kg solvent}} = \frac{50.0\ \text{moles KOH}}{3.4\ \text{kg water}} = 15\ \text{m}$$

The molality is 15 m.

7. Molality is number of moles of solute divided by kg of solvent. So we have to get our units into moles and kg:

$$\frac{35.0\ \cancel{\text{g CaBr}_2}}{1} \times \frac{1\ \text{mole CaBr}_2}{199.9\ \cancel{\text{g CaBr}_2}} = 0.175\ \text{moles CaBr}_2$$

$$\frac{657 \text{ g}}{1} \times \frac{1 \text{ kg}}{1{,}000 \text{ g}} = 0.657 \text{ kg}$$

$$\text{molality} = \frac{\#\text{ moles solute}}{\#\text{ kg solvent}} = \frac{0.175 \text{ moles CaBr}_2}{0.657 \text{ kg water}} = 0.266 \text{ m}$$

The molality is 0.266 m.

8. Freezing point depression is determined by Equation (11.2). We are already given two of the four variables in the equation (K_f,ΔT) , and we can calculate a third (i). Molality is the only unknown, so we can solve for it. NaCl, since it is ionic, will split up into ions (one sodium ion and one chloride ion), so i= 2:

$$\Delta T = -i \cdot K_f \cdot m$$

$$-15\ ^\circ\text{C} = -2 \cdot 1.86 \frac{^\circ\text{C}}{\text{molal}} \cdot \text{m}$$

$$\text{m} = \frac{-15\ ^\circ\text{C}}{-2 \cdot 1.86 \frac{^\circ\text{C}}{\text{molal}}} = 4.0 \text{ molal}$$

The molality is 4.0 m.

9. In a freezing point depression problem, we must use Equation (11.2). However, in order to use that equation, we must know K_f, i, and m. Right now, we only know K_f. However, we have been given enough information to calculate both "i" and "m." To calculate m:

$$\frac{35.0 \text{ g NH}_3}{1} \times \frac{1 \text{ mole NH}_3}{17.0 \text{ g NH}_3} = 2.06 \text{ moles NH}_3$$

$$\frac{350.0 \text{ g}}{1} \times \frac{1 \text{ kg}}{1{,}000 \text{ g}} = 0.3500 \text{ kg}$$

$$\text{m} = \frac{2.06 \text{ moles NH}_3}{0.3500 \text{ kg water}} = 5.89 \text{ m}$$

To figure out "i", we just have to think about how NH_3 dissolves. Being a polar covalent compound, each molecule dissolves individually, so i = 1.

Now that we have all of the components of Equation (11.2), we can use it:

$$\Delta T = -i \cdot K_f \cdot m = -1 \cdot 1.86 \frac{^\circ\text{C}}{\text{m}} \cdot 5.89 \text{ m} = -11.0\ ^\circ\text{C}$$

Our answer indicates that the freezing point of this solution is 11.0 °C lower than that of water. Thus, the answer to our problem is that the freezing point of the solution is -11.0 °C.

10. To calculate boiling points, we must use Equation (11.3). To do that, however, we must know "i" and "m". To calculate "m":

$$\frac{100.0\ \cancel{\text{g ZnCl}_2}}{1} \times \frac{1\ \text{mole ZnCl}_2}{136.4\ \cancel{\text{g ZnCl}_2}} = 0.7331\ \text{moles ZnCl}_2$$

$$\frac{750.0\ \cancel{\text{g}}}{1} \times \frac{1\ \text{kg}}{1{,}000\ \cancel{\text{g}}} = 0.7500\ \text{kg}$$

$$\text{m} = \frac{0.7331\ \text{moles ZnCl}_2}{0.7500\ \text{kg water}} = 0.9775\ \text{m}$$

Since zinc(II) chloride is an ionic compound, it dissolves by splitting up into its one zinc ion and its two chloride ions. Thus, i = 3.

$$\Delta T = i \cdot K_b \cdot m = 3 \cdot 0.512 \frac{^\circ\text{C}}{\cancel{\text{m}}} \cdot 0.9775\ \cancel{\text{m}} = 1.50\ ^\circ\text{C}$$

This means that the boiling point of the solution is 1.50 °C *higher* than that of pure water. The boiling point of pure water is 100.0 °C, so the boiling point of this solution is 101.5 °C.

SOLUTIONS TO THE PRACTICE PROBLEMS FOR MODULE #12

1. This problem asks you to predict how a gas will change when you change some of the conditions under which it is stored. This means that you need to use the combined gas law (Equation 12.10).

$$\frac{P_1V_1}{T_1} = \frac{P_2V_2}{T_2}$$

According to this problem, $P_1 = 755$ torr, $V_1 = 1.05$ L, and $P_2 = 625$ torr. Also, the problem states that the temperature does not change, thus T_1 and T_2 cancel out:

$$\frac{P_1V_1}{\cancel{T_1}} = \frac{P_2V_2}{\cancel{T_2}}$$

We can now rearrange the equation to solve for the new volume:

$$\frac{P_1V_1}{P_2} = V_2$$

Now we can put in the numbers and determine the new volume:

$$\frac{755\ \cancel{\text{mmHg}} \cdot 1.05\ \text{L}}{625\ \cancel{\text{mmHg}}} = 1.27\ \text{L}$$

Thus, the balloon expands to 1.27 liters.

2. This is obviously another combined gas law problem, with $P_1 = 780$ torr, $V_1 = 46.1$ mL, $T_1 = 25\ ^{\circ}C$, $P_2 = 1.00$ atm (standard pressure), and $T_2 = 273$ K (standard temperature). The problem asks us to determine the new volume, so we have to rearrange Equation (12.10) to solve for V_2:

$$\frac{P_1V_1T_2}{T_1P_2} = V_2$$

Before we can plug in the numbers, however, we need to convert T_1 to Kelvin.

$$T_1 = 25 + 273.15 = 298\ \text{K}$$

Additionally, we need to make the pressure units the same. We can do this by converting torr into atm or vice-versa. I will choose to do the latter:

$$P_2 = \frac{1.00\ \cancel{\text{atm}}}{1} \times \frac{760\ \text{torr}}{1\ \cancel{\text{atm}}} = 7.60 \times 10^2\ \text{torr}$$

Now we can plug in the numbers:

$$\frac{780\ \text{torr} \cdot 46.1\ \text{mL} \cdot 273\ \text{K}}{298\ \text{K} \cdot 7.60 \times 10^2\ \text{torr}} = V_2$$

$$43\ \text{mL} = V_2$$

So the volume increases to 43 mL.

3. In this combined gas law problem, we are asked to calculate the pressure of a container of gas under conditions of constant volume. Thus, we need to rearrange the equation to solve for P_2, realizing that the volumes cancel because they are the same:

$$\frac{P_1 V_1}{T_1} = \frac{P_2 V_2}{T_2}$$

$$\frac{P_1 T_2}{T_1} = P_2$$

To be able to calculate the new volume, we first need to get the temperatures into Kelvin:

$$T_1 = 20.0 + 273.15 = 293.2\ \text{K}$$

$$T_2 = 55.0 + 273.15 = 328.2\ \text{K}$$

Now we can plug in our numbers:

$$\frac{123.0\ \text{atms} \cdot 328.2\ \text{K}}{293.2\ \text{K}} = 137.7\ \text{atms}$$

Thus, the pressure in the container after it warms up is 137.7 atms.

4. This problem requires you to recognize that when a gas is collected over water, the gas is contaminated with water vapor. Thus, the 787 torr of gas is the total pressure of carbon dioxide *plus* water vapor. To determine the partial pressure of water vapor, we need only look at Table 12.1. At 27 °C, the vapor pressure of water is 26.7 torr. Thus, Dalton's Law becomes:

$$P_T = P_{\text{carbon dioxide}} + P_{\text{water vapor}}$$

$$787\ \text{torr} = P_{\text{carbon dioide}} + 26.7\ \text{torr}$$

$$P_{\text{carbon dioxide}} = 787\ \text{torr} - 26.7\ \text{torr} = 7.60 \times 10^2\ \text{torr}$$

Thus, only 7.60×10^2 torr of carbon dioxide was collected.

5. Mole fraction is defined as the number of *moles* of component divided by the total number of moles. Right now, the problem gives us *grams*, not moles. Thus, we must first convert from grams to moles:

$$\frac{15.0\ \cancel{g\,SO_2}}{1} \times \frac{1\ mole\,SO_2}{64.1\ \cancel{g\,SO_2}} = 0.234\ moles\,SO_2$$

$$\frac{12.1\ \cancel{g\,NO}}{1} \times \frac{1\ mole\,NO}{30.0\ \cancel{g\,NO}} = 0.403\ moles\,NO$$

$$\frac{2.5\ \cancel{g\,SO_3}}{1} \times \frac{1\ mole\,SO_3}{80.1\ \cancel{g\,SO_3}} = 0.031\ moles\,SO_3$$

Now that we have the number of moles of each component, we can calculate the total number of moles in the mixture:

$$\text{Total number of moles} = 0.234\ \text{moles} + 0.403\ \text{moles} + 0.031\ \text{moles} = 0.668\ \text{moles}$$

Plugging that into Equation (12.12):

$$X_{SO_2} = \frac{0.234\ \cancel{moles}}{0.668\ \cancel{moles}} = 0.350$$

$$X_{NO} = \frac{0.403\ \cancel{moles}}{0.668\ \cancel{moles}} = 0.603$$

$$X_{SO_3} = \frac{0.031\ \cancel{moles}}{0.668\ \cancel{moles}} = 0.046$$

The mole fractions of SO_2, NO, and SO_3 are 0.350, 0.603, and 0.046, respectively.

6. Using the mole fractions we just obtained, this problem is an easy application of Equation (12.13)

$$P_1 = X_1 \cdot P_T$$

$$P_{NO} = 0.603 \cdot 1.1\ atm = 0.66\ atm$$

$$P_{SO_2} = 0.350 \cdot 1.1\ atm = 0.39\ atm$$

$$P_{SO_3} = 0.046 \cdot 1.1\ atm = 0.051\ atm$$

The partial pressures of NO, SO_2, and SO_3 are 0.66 atm, 0.39 atm, and 0.051 atm, respectively.

7. In this problem, we are given the partial pressure of each gas. By Dalton's Law, the total pressure is just the sum of the individual pressures:

$$P_T = 5.00 \text{ atms} + 2.00 \text{ atms} + 0.50 \text{ atms} = 7.50 \text{ atms}$$

By Equation (12.13), then, we can calculate the mole fractions:

$$P_1 = X_1 \cdot P_T$$

$$X_1 = \frac{P_1}{P_T}$$

Plugging the numbers in for each gas:

$$X_{N_2} = \frac{5.00 \cancel{\text{atms}}}{7.50 \cancel{\text{atms}}} = 0.667$$

$$X_{O_2} = \frac{2.00 \cancel{\text{atms}}}{7.50 \cancel{\text{atms}}} = 0.267$$

$$X_{Ar} = \frac{0.50 \cancel{\text{atms}}}{7.50 \cancel{\text{atms}}} = 0.067$$

The mole fractions of N_2, O_2, and Ar are 0.667, 0.267, and 0.067, respectively.

8. In this problem, we are given pressure and temperature and the number of moles. We are then asked to calculate V. We can do this by rearranging the ideal gas law:

$$PV = nRT$$

$$V = \frac{nRT}{P}$$

We know that $R = 0.0821 \frac{L \cdot atm}{mole \cdot K}$, so in order for the equation to work, our temperature must be converted into 298 K. Also, we do not have n yet. We do, however, have mass, so we can convert it into moles :

$$\frac{16.1 \cancel{\text{g Cl}_2}}{1} \times \frac{1 \text{ mole Cl}_2}{71.0 \cancel{\text{g Cl}_2}} = 0.227 \text{ moles Cl}_2$$

Now that we have all of the correct units, we can plug the numbers into the equation.

$$V = \frac{0.227\ \cancel{moles} \cdot 0.0821 \frac{L \cdot \cancel{atm}}{\cancel{mole} \cdot \cancel{K}} \cdot 298\ \cancel{K}}{1.00\ \cancel{atm}} = 5.55\ L$$

The volume is 5.55 liters.

9. In this problem, we are given the amount of limiting reactant and asked to determine how much of another reactant is used. The problem here is that the amount of limiting reactant is not given in grams or moles. Instead it is given in P, V, and T. Thus, we must use the ideal gas law to determine the number of moles of limiting reactant. Before we can do that, however, volume must be converted to liters to make it consistent with the volume unit in R.

$$\frac{56.7\ \cancel{mL}}{1} \times \frac{0.001\ L}{1\ \cancel{mL}} = 0.0567\ L$$

We also must convert the temperature to 295 K. Now we can plug the numbers into the equation:

$$n = \frac{PV}{RT} = \frac{1.23\ \cancel{atm} \cdot 0.0567\ \cancel{L}}{0.0821 \frac{\cancel{L} \cdot \cancel{atm}}{mole \cdot \cancel{K}} \cdot 295\ \cancel{K}} = 0.00288\ moles$$

Now that we have the moles of limiting reactant, this becomes a stoichiometry problem:

$$\frac{0.00288\ \cancel{moles\ H_2S}}{1} \times \frac{4\ moles\ Ag}{2\ \cancel{moles\ H_2S}} = 0.00576\ moles\ Ag$$

$$\frac{0.00576\ \cancel{moles\ Ag}}{1} \times \frac{107.9\ g\ Ag}{1\ \cancel{mole\ Ag}} = 0.622\ g\ Ag$$

Therefore, 0.622 g of silver will tarnish.

10. In this stoichiometry problem, we are given the amount of limiting reactant and asked to calculate how much product will be made. We start by converting the amount of limiting reactant to moles:

$$\frac{1500.0\ \cancel{g\ H_2O_2}}{1} \times \frac{1\ mole\ H_2O_2}{34.0\ \cancel{g\ H_2O_2}} = 44.1\ moles\ H_2O_2$$

We can then use stoichiometry to determine the number of moles of H_2O produced:

$$\frac{44.1\ \cancel{moles\ H_2O_2}}{1} \times \frac{8\ moles\ H_2O}{7\ \cancel{moles\ H_2O_2}} = 50.4\ moles\ H_2O$$

Now we need to use the ideal gas law, realizing that we must convert the temperature to Kelvin:

$$PV = nRT$$

$$V = \frac{nRT}{P} = \frac{50.4\ \text{moles} \cdot 0.0821 \frac{\text{L} \cdot \text{atm}}{\text{mole} \cdot \text{K}} \cdot 810\ \text{K}}{1.2\ \text{atm}} = 2.8 \times 10^3\ \text{L}$$

The volume produced is 2.8×10^3 L.

SAMPLE CALCULATIONS FOR EXPERIMENT 12.1

Mass of Vinegar: 2.00×10^2 g
Since your mass scale is most likely marked of in units of 10 g, the most precision you have is to the ones place. Thus, scientific notation is the only way you can report this answer.
Circumference of the balloon: 6.158 dm

Radius of the balloon: $\frac{6.158\ \text{dm}}{2\pi} = \frac{6.158\ \text{dm}}{2 \cdot (3.1416)} = 0.9801\ \text{dm}$

Volume of the balloon: $\frac{4}{3}\pi r^3 = \frac{4}{3} \cdot (3.1416) \cdot (0.9801\ \text{dm})^3 = 3.944\text{dm}^3 = 3.944\ \text{L}$

Atmospheric Pressure: 0.945 atm
Temperature: 23.8 °C = 297.0 K
Water Vapor Pressure: 22.4 torr = 0.0295 atm

Pressure of CO_2 Formed: 0.945 atm - 0.0295 atm = 0.916 atm

Now we can get the moles of CO_2 made in the reaction:

$$n = \frac{PV}{RT} = \frac{0.916\ \text{atm} \cdot 3.944\ \text{L}}{0.0821 \frac{\text{L} \cdot \text{atm}}{\text{mole} \cdot \text{K}} \cdot 297.0\ \text{K}} = 0.148\ \text{moles}$$

Now we can use stoichiometry to determine the number of moles of acetic acid:

$$\frac{0.148\ \text{moles}\ CO_2}{1} \times \frac{1\ \text{mole}\ C_2H_4O_2}{1\ \text{mole}\ CO_2} = 0.148\ \text{moles}\ C_2H_4O_2$$

The number of grams of acetic acid used in the experiment, then, is:

$$\frac{0.148\ \text{moles}\ C_2H_4O_2}{1} \times \frac{60.0\ \text{g}\ C_2H_4O_2}{1\ \text{mole}\ C_2H_4O_2} = 8.88\ \text{g}\ C_2H_4O_2$$

Since I used 200.0 g initially, this means that 4.44% of the vinegar was acid.

SOLUTIONS TO THE PRACTICE PROBLEMS FOR MODULE #13

1. The chemical equation contains no phases, so we must use bond energies to solve this problem. This means we start with Lewis structures:

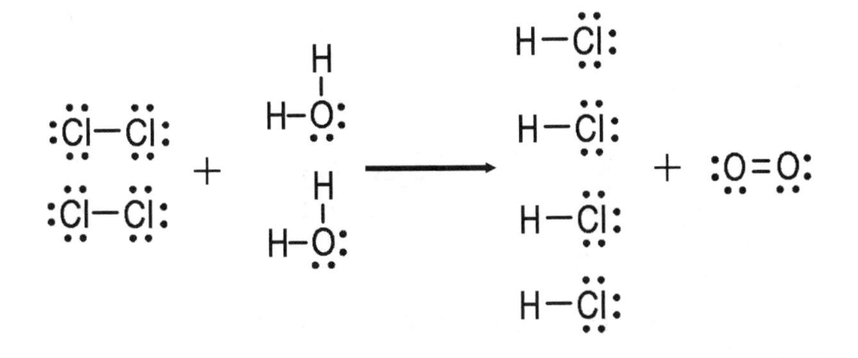

Using this picture and Table 13.1, we know that two Cl-Cl bonds (each worth 240 kJ/mole) and four H-O bonds (each worth 459 kJ/mole) must be broken while four H-Cl bonds (each worth 428 kJ/mole) and one O=O bond (worth 494 kJ/mole) must be formed. Equation (13.6) becomes:

$$\Delta H = (2\ \cancel{\text{moles}}) \times (240 \frac{\text{kJ}}{\cancel{\text{mole}}}) + (4\ \cancel{\text{moles}}) \times (459 \frac{\text{kJ}}{\cancel{\text{mole}}}) - (4\ \cancel{\text{moles}}) \times (428 \frac{\text{kJ}}{\cancel{\text{mole}}}) - (1\ \cancel{\text{mole}}) \times (494 \frac{\text{kJ}}{\cancel{\text{mole}}})$$

$$\Delta H = \underline{110\ \text{kJ}}$$

Since we are adding and subtracting here, we use the rule of addition and subtraction. That means we worry about decimal place. The number 240 has its last significant figure in the tens place, so that's as precise as the answer can be.

2. Since we know the phases for the substances in a combustion reaction, we can us Hess's Law here. The definition of combustion tells us that the reaction is:

$$2C_6H_6\ (l) + 15O_2\ (g) \rightarrow 12CO_2\ (g) + 6H_2O\ (g)$$

The ΔH_f^o of O_2 (g) is zero since it is oxygen's elemental form. The rest of the ΔH_f^o's are either given in the problem or are in the Table 13.2, so we can apply Equation (13.9) directly:

$$\Delta H^o = (12\ \cancel{\text{moles}}) \times (-394 \frac{\text{kJ}}{\cancel{\text{mole}}}) + (6\ \cancel{\text{moles}}) \times (-242 \frac{\text{kJ}}{\cancel{\text{mole}}}) - (2\ \cancel{\text{moles}}) \times (49.00 \frac{\text{kJ}}{\cancel{\text{mole}}})$$

$$\Delta H^o = \underline{-6{,}278\ \text{kJ}}$$

Using the rule of addition and subtraction, we can report our answer to the ones place, since the least precise bond energies have their last significant figure in the ones place.

3. In this problem, we are given the ΔH^o of the reaction, and we are asked to calculate the ΔH_f^o of one of the reactants. This isn't hard, however, because Equation (13.9) relates the ΔH^o of the equation to

the ΔH_f^o's of all of the substances in the equation. Since we know that the ΔH_f^o of O_2 (g) is zero and the ΔH_f^o of SO_3 (g) is in Table 13.2, the ΔH_f^o of SO_2 (g) is our only unknown in Equation (13.9):

$$-198 \text{ kJ} = (2 \text{ moles}) \times (-396 \frac{\text{kJ}}{\text{mole}}) - (2 \text{ moles}) \times (\Delta H_f^o \text{ of } SO_2)$$

Rearranging the equation gives us:

$$\Delta H_f^o \text{ of } SO_2 = \frac{-198 \text{ kJ} - (2 \cancel{\text{moles}}) \times (-396 \frac{\text{kJ}}{\cancel{\text{mole}}})}{-2 \text{ moles}} = -297 \frac{\text{kJ}}{\text{mole}}$$

The standard enthalpy of formation for gaseous sulfur dioxide is -297 kJ/mole.

4. In #2, the answer was ΔH = -6,278 kJ/mole. This means that energy is a product in this reaction:

$$2C_6H_6 \text{ (l)} + 15O_2 \text{ (g)} \rightarrow 12CO_2 \text{ (g)} + 6H_2O \text{ (g)} + 6{,}278 \text{ kJ}$$

Now we can do stoichiometry:

$$\frac{500.0 \cancel{\text{g } C_6H_6}}{1} \times \frac{1 \text{ mole } C_6H_6}{78.1 \cancel{\text{g } C_6H_6}} = 6.40 \text{ moles } C_6H_6$$

$$\frac{6.40 \cancel{\text{moles } C_6H_6}}{1} \times \frac{6{,}278 \text{ kJ}}{2 \cancel{\text{moles } C_6H_6}} = \underline{20{,}100 \text{ kJ}}$$

5. In order to answer any question about the energy involved in a chemical reaction, we must first get ΔH. Since Table 13.2 contains all the information we need for this problem, we will use Hess's Law to do this calculation:

$$\Delta H^o = (2 \cancel{\text{moles}}) \times (-45.9 \frac{\text{kJ}}{\cancel{\text{mole}}}) + (4 \cancel{\text{moles}}) \times (-242 \frac{\text{kJ}}{\cancel{\text{mole}}}) - (2 \cancel{\text{moles}}) \times (33.2 \frac{\text{kJ}}{\cancel{\text{mole}}})$$

$$\Delta H^o = -1{,}126 \text{ kJ}$$

This means that energy is a product in the reaction:

$$2NO_2 \text{ (g)} + 7H_2 \text{ (g)} \rightarrow 2NH_3 \text{ (g)} + 4H_2O \text{ (g)} + 1{,}126 \text{ kJ}$$

Now we can do stoichiometry:

$$\frac{35.0 \cancel{\text{g } NH_3}}{1} \times \frac{1 \text{ mole } NH_3}{17.0 \cancel{\text{g } NH_3}} = 2.06 \text{ moles } NH_3$$

$$\frac{2.06\ \cancel{\text{moles NH}_3}}{1} \times \frac{1{,}126\ \text{kJ}}{2\ \cancel{\text{moles NH}_3}} = 1{,}160\ \text{kJ}$$

Thus, when 35.0 grams of ammonia are made this way, 1,160 kJ of energy are produced.

6. a. In this reaction, there is one molecule of gas on the products side and no molecules of gas on the reactants side. Thus, the products have a higher entropy. This means ΔS is positive.

b. In this reaction, there are no molecules of gas on the products side and two molecules of gas on the reactants side. This means that the reactants are more disordered than the products, so ΔS is negative.

c. In this reaction, there are two molecules of gas on the products side and four molecules of gas on the reactants side. This means that the reactants are more disordered than the products, so ΔS is negative.

7. Table 13.3 lists all of the substances in this reaction, so we can use it and Equation (13.12) to solve this problem:

$$\Delta S^\circ = (1\ \cancel{\text{mole}}) \times (76.1 \frac{\text{J}}{\cancel{\text{mole}} \cdot \text{K}}) - (1\ \cancel{\text{mole}}) \times (69.9 \frac{\text{J}}{\cancel{\text{mole}} \cdot \text{K}}) - (1\ \cancel{\text{mole}}) \times (38.2 \frac{\text{J}}{\cancel{\text{mole}} \cdot \text{K}}) = -32.0 \frac{\text{J}}{\text{K}}$$

8. This problem gives us ΔH and ΔS and asks us at what temperature the reaction is spontaneous. In other words, we need to see for what temperatures ΔG is negative. Thus:

$$\Delta H - T\Delta S < 0$$

In order to use this equation, though, we need to get our units consistent:

$$\frac{-324\ \cancel{\text{J}}}{\text{mole} \cdot \text{K}} \times \frac{1\ \text{kJ}}{1{,}000\ \cancel{\text{J}}} = -0.324 \frac{\text{kJ}}{\text{mole} \cdot \text{K}}$$

Now we can use the equation:

$$-1{,}023 \frac{\text{kJ}}{\text{mole}} - T \cdot (-0.324 \frac{\text{kJ}}{\text{mole} \cdot \text{K}}) < 0$$

We can solve this equation like any algebraic equation. We just need to remember that if we divide or multiply by a negative number, we must reverse the inequality sign:

$$T \cdot (0.324 \frac{\text{kJ}}{\text{mole} \cdot \text{K}}) < 1{,}023 \frac{\text{kJ}}{\text{mole}}$$

$$T < \frac{1023 \frac{\cancel{\text{kJ}}}{\cancel{\text{mole}}}}{0.324 \frac{\cancel{\text{kJ}}}{\cancel{\text{mole}} \cdot \text{K}}}$$

$$T < 3.16 \times 10^3\ \text{K}$$

This reaction, then, is spontaneous for all temperatures lower than 3.16×10^3 K.

9. The ΔH of a reaction is the difference in energy between the products and the reactants. According to this diagram, the products look to be at an energy of 1200 kJ while the reactants look to be at an energy of 200 kJ. This means that ΔH = 1200 kJ - 200 kJ = 1000 kJ. The activation energy is defined as the difference in energy between the intermediate phase and the reactants. Thus, the activation energy = 1500 - 200 = 1300 kJ. The reaction is endothermic.

10. This reaction is at 298 K, so we can use Table 13.4 and Equation (13.14) to calculate ΔG:

$$\Delta G^{\circ} = (6\ \cancel{\text{moles}}) \times (-175 \frac{\text{kJ}}{\cancel{\text{mole}}}) - (2\ \cancel{\text{moles}}) \times (125 \frac{\text{kJ}}{\cancel{\text{mole}}}) = -1.300 \times 10^{3}\ \text{kJ}$$

Since ΔG is negative, the reaction is spontaneous at 298 K.

SAMPLE CALCULATIONS FOR EXPERIMENT 13.1

Mass of 10 Tablespoons of Lye: 212 g

Mass of one Teaspoon of Lye: $\frac{212\ \text{g}}{30} = 7.07\ \text{g}$

The 30 is exact, as it represents an integer number of teaspoons. You did not measure the mass of 9.9 tablespoons or 10.1 tablespoons of lye. You measured the mass of *exactly* 10 tablespoons.

Mass of the Lye used in the Experiment: $\frac{7.07\ \text{g}}{2} = 3.54\ \text{g}$

Initial temperature: 24.8 °C
Final temperature: 28.4 °C

Mass of calorimeter contents: $100.0\ \cancel{\text{mL}} \times 0.99 \frac{\text{g}}{\cancel{\text{mL}}} + 3.54\ \text{g} = 103\ \text{g}$

When you take 100.0 times 0.99, your significant figures allow you to report the answer as 99 grams. When you add 3.54 g to that, the rule of addition and subtraction allows you to report your answer to the ones place.

Change in temperature: ΔT = 28.4 °C - 24.8 °C = 3.6 °C

$$q = m \cdot c \cdot \Delta T = (103\ \cancel{\text{g}}) \cdot \left(4.1 \frac{\text{J}}{\cancel{\text{g}} \cdot \cancel{^{\circ}\text{C}}}\right) \cdot (3.6\ \cancel{^{\circ}\text{C}}) = 1{,}500\ \text{J}$$

One molecule of NaOH has a mass of 40.0 amu, which means 1 mole NaOH = 40.0 g NaOH:

$$\text{Moles NaOH} = \frac{3.54\ \cancel{\text{g NaOH}}}{1} \times \frac{1\ \text{mole NaOH}}{40.0\ \cancel{\text{g NaOH}}} = 0.0885\ \text{moles NaOH}$$

$$\Delta H = \frac{1{,}500\ \text{J}}{0.0885\ \text{moles}} = 17{,}000 \frac{\text{J}}{\text{mole}}$$

NOTE: Your answer might be quite different than mine. That's fine. It depends a lot on the brand of lye that you used.

SOLUTIONS TO THE PRACTICE PROBLEMS FOR MODULE #14

1. Since the souring of milk is governed by a chemical reaction, storing the milk in the refrigerator will cause the reaction to run at a lower temperature. A lower temperature slows down the reaction that causes souring, making the milk last longer.

2. The rate equation for this reaction will look like:

$$R = k[NO]^x[O_2]^y$$

To figure out k, x, and y, we have to look at the data from the experiment. The value for x can be determined by comparing two trials in which the concentration of NO changes, but the concentration of O_2 stays the same. This would correspond to trials 1 and 2. In these two trials, the concentration of NO doubled, and the rate went up by a factor of 4. This means that x = 2, because the only way you can get a 4-fold increase in rate from a doubling of the concentration is by squaring the concentration. The value for y can be determined by looking at trials 1 and 3, where the concentration of NO stayed the same but the concentration of O_2 doubled. When that happened, the rate increased by a factor of 2. Since rate doubled when concentration doubled, that means y = 1. Thus, the rate equation becomes:

$$R = k[NO]^2[O_2]$$

Now that we have x and y, we only need to find out the value for k. We can do this by using any one of the trials in the experiment and plugging the data into the equation. The only unknown will be k, and we can therefore solve for it:

$$R = k[NO]^2[O_2]$$

$$0.0281\,\frac{M}{s} = k \cdot (0.0125\ M)^2 \cdot (0.0253\ M)$$

$$k = \frac{0.0281\,\frac{\cancel{M}}{s}}{(0.0125\ M)^2 \cdot (0.0253\ \cancel{M})} = 7.11 \times 10^3\,\frac{1}{M^2 \cdot s}$$

Thus, the final rate equation is:

$$\underline{R = (7.11 \times 10^3\,\frac{1}{M^2 \cdot s}) \cdot [NO]^2[O_2]}$$

3. The rate equation will take on the form:

$$R = k[C_4Cl_9OH]^x[F^-]^y$$

To determine x and y, we look at trials where the concentration of one reactant stayed the same and the concentration of the other reactant changed. In trials 1 and 2, the concentration of C_4Cl_9OH remained the same but the concentration of F^- doubled. When that happened, the rate stayed the same. This means that y = 0, because the rate is not affected by the change in concentration of F^-. In the same way, between trials 2 and 3, the F^- concentration remained constant, but the C_4Cl_9OH concentration doubled. When that happened, the rate doubled. This means x = 1. The rate equation, then, looks like:

$$R = k[C_4Cl_9OH][F^-]^0 \text{ or } R = k[C_4Cl_9OH]$$

To solve for k, we can use the data from any trial and plug it into our rate equation. We can then solve for k:

$$R = k[C_4Cl_9OH]$$

$$0.0202\ \frac{M}{s} = k \cdot (0.25\ M)$$

$$k = \frac{0.0202\ \frac{\not{M}}{s}}{(0.25\ \not{M})} = 0.081\ \frac{1}{s}$$

The overall rate equation, then is

$$\underline{R = (0.081\frac{1}{s}) \cdot [C_4Cl_9OH]}$$

4. Second order with respect to I_2 means that the exponent in the rate equation that is attached to I_2 equals 2. In the same way, first order in Br_2 means that the exponent tied to that reactant is 1. The overall order, therefore is 3. Since the rate constant is given, we can say that the rate equation is:

$$R = (1.1 \times 10^{-3}\ \frac{1}{M^2 \cdot s}) \cdot [I_2]^2[Br_2]$$

To determine the rate, then, we just need to plug the concentrations into the equation:

$$R = (1.1 \times 10^{-3}\ \frac{1}{\not{M}^2 \cdot s}) \cdot (0.5\ \not{M})^2 (0.5\ M) = \underline{1 \times 10^{-4}\ \frac{M}{s}}$$

5. The rate of a chemical reaction doubles for every 10 °C that the temperature is raised. Alternatively, the rate decreases by a factor of 2 for every 10 °C that the temperature is lowered. In this problem, the temperature is lowered for 5 ten degree increments. This means that the old rate must be divided by 2 for a total of five times:

$$R = 0.0167\ \frac{M}{s} \div 2 \div 2 \div 2 \div 2 \div 2 = 0.000522\ \frac{M}{s}$$

The rate at the lower temperature is 0.000522 M/s.

6. Since chemical reaction rate doubles for every 10 °C increment, then to increase the rate of the reaction by a factor of 16, I just need to raise the temperature by 4 ten degree increments. That way, I will multiply the old rate by 2x2x2x2, which equals 16. Thus, to increase the reaction rate by a factor

of 16, I just raise the temperature by 40 degrees. Therefore, the new temperature should be 25 °C + 40°C = 65 °C.

7. The rate constant increases with increasing temperature, so that fact alone rules out graphs I and II. We also know, however, that the rate constant increases dramatically with temperature. This rules out graph IV. Graph III looks like the one presented in the module, so it is the correct answer.

8. This is a heterogeneous catalyst, because it is in a different phase than the reactants.

9. A catalyst lowers the activation energy, which increases the rate of the reaction. The better the catalyst, the lower the activation energy and hence the faster the reaction. Thus, the diagram with the largest activation energy would correspond to the reaction without a catalyst; the one with the lowest activation energy would correspond to the reaction with the best catalyst, and the one with a medium activation energy would correspond to the reaction with the other catalyst. Thus, III represents the reaction with no catalyst, I represents the reaction with the catalyst that speeds up the rate 3x, and II represents the reaction with the catalyst that speeds up rate 10x.

10. A catalyst is a substance that does not get used up in the reaction. Thus, in a reaction mechanism, a catalyst must be a reactant in the first step of the mechanism and a product in the last step. Thus, Cl is the catalyst here. Since it is in the same phase as the reactants, it is a homogeneous catalyst.

SOLUTIONS TO THE PRACTICE PROBLEMS FOR MODULE #15

1. According to Equation (15.2), the equilibrium constant for this reaction is:

$$K = \frac{[N_2]_{eq}[H_2]_{eq}^3}{[NH_3]_{eq}^2}$$

Plugging those equilibrium concentrations into the equation:

$$K = \frac{(0.45\ M)(0.63\ M)^3}{(0.24\ M)^2} = 2.0\ M^2$$

The equilibrium constant is $\underline{2.0\ M^2}$.

2. To get the equation for the equilibrium constant, we have to realize that we ignore both $Pb(NO_3)_2$ and PbO, because they are solids. Thus, Equation (15.2) becomes:

$$K = [NO_2]_{eq}^4[O_2]_{eq}$$

Since we ignore solids, we ignore the amounts of both given in the problem. We only use the equilibrium concentrations of the gases. Plugging those equilibrium concentrations into the equation:

$$K = (0.18\ M)^4(0.045\ M) = 4.7 \times 10^{-5}\ M^5$$

The equilibrium is $\underline{4.7 \text{ x } 10^{-5}\ M^5}$.

3. a. When the equilibrium constant is very small, only reactants are present at any significant concentration once equilibrium is achieved. Thus, this reaction can be written as its reverse:

$$2NOCl\ (g) + O_2\ (g) \leftarrow 2NO_2\ (g) + Cl_2\ (g)$$

To make it a bit easier to read, we usually turn it around:

$$\underline{2NO_2\ (g) + Cl_2\ (g) \rightarrow 2NOCl\ (g) + O_2\ (g)}$$

b. This equilibrium constant is neither large nor small, so we cannot replace the double arrow with a single arrow.

c. When the equilibrium constant is very large, there is only product present in significant amounts when the reaction reaches equilibrium. Thus, this reaction can be written with a single arrow:

$$\underline{2NO_2 \rightarrow N_2O_4}$$

4. In this reaction, we must ignore water because it is a liquid. The equation for the equilibrium constant, then, is:

$$K = \frac{[CO_2]_{eq}}{[H_2CO_3]_{eq}}$$

If the concentrations are, in fact, equilibrium concentrations, then the equation should equal the value given for K.

$$K = \frac{(3.5\ \cancel{M})}{(2.3\ \cancel{M})} = 1.5$$

This is nowhere near the value of 2.3×10^4. Since our calculated value is too low, K must get bigger; thus, the reaction must shift towards products.

5. The equation for the equilibrium constant here is:

$$K = \frac{[SO_3]_{eq}^{\ 2}}{[SO_2]_{eq}^{\ 2}[O_2]_{eq}}$$

If the concentrations are, in fact, equilibrium concentrations, then the equation should equal the value given for K.

$$K = \frac{(0.280\ \cancel{M})^2}{(0.340\cancel{M})^2(0.154\ M)} = 4.40\ \frac{1}{M}$$

This is equal to the value for K, so the reaction is at equilibrium.

6. a. We ignore solids as a source of stress on the equilibrium, so nothing will happen.

b. We also ignore liquids a source of stress on the equilibrium, so nothing will happen.

c. When H_2CO_3 is removed, the reverse reaction will slow down. This will make the forward reaction faster than the reverse, causing a shift towards the products. Thus, more CaO will be made. Remember, solids and liquids are not sources of stress because their *concentrations* do not change. However, if we stress the equilibrium by varying the amount of another substance in the equation, the *amounts* of liquids and solids can change.

7. a. The reaction is endothermic, which means energy is a reactant. If the temperature is raised, the equilibrium shifts away from the side with the energy, so the H_2 concentration will go down.

b. When temperature is lowered, the reaction shifts towards the side with energy. Thus, the concentration of HF will lower.

8. a. This is an exothermic reaction, so energy is a product. When the temperature is raised, the reaction shifts away from the side with the energy, so the concentration of H_2 will increase.

b. When pressure is raised, the reaction shifts away from the side with the most gas molecules. There are four gas molecules on the reactants side and only two on the products side. Thus, the reaction will shift away from reactants, causing <u>the concentration of NH_3 to rise</u>.

c. When pressure is lowered, the reaction shifts towards the side with more gas molecules, making <u>the concentration of N_2 increase</u>.

9. The ionization constant is simply the equilibrium constant for the acid ionization reaction. In order to determine the ionization reaction, you simply take the acid in its aqueous phase and remove an H^+. When you remove an H^+ from HCN, you are left with CN^-. In the end, then, the aqueous acid is the reactant, and the H^+ and CN^- (both in aqueous phase) will be the products:

$$HCN\ (aq) \leftrightharpoons H^+\ (aq) + CN^-\ (aq)$$

The equilibrium constant for this reaction is the ionization constant, K_a:

$$\underline{K_a = \frac{[H^+][CN^-]}{[HCN]}}$$

10. The base ionization reaction involves having the base accept an H^+ ion from water. If SO_4^{2-} accepts an H^+, it becomes HSO_4^-. When water gives up that H^+, it becomes OH^-. All of this takes place in water, so the water phase is liquid and everything else is aqueous:

$$SO_4^{2-}\ (aq) + H_2O\ (l) \leftrightharpoons HSO_4^-\ (aq) + OH^-\ (aq)$$

The ionization constant, then, is just the equilibrium constant of this reaction:

$$\underline{K_b = \frac{[OH^-][HSO_4^-]}{[SO_4^{2-}]}}$$

SOLUTIONS TO THE PRACTICE PROBLEMS FOR MODULE #16

1. When a substance is made up of only one type of atom, the oxidation number of the atom is the charge of the substance divided by the number of atoms in the substance.

a. <u>0</u> b. <u>+2</u> c. <u>0</u> d. <u>0</u> e. <u>-2</u>

2. a. Rule #6 says that O will have an oxidation number of -2. Since all of the oxidation numbers must add up to the overall charge, the Mn is +4. The oxidation numbers, then, are: <u>Mn: +4, O: -2</u>

b. Rule #5 tells us that H is +1 here. Rule number #6 tells us that O is -2. From H and O, we have a sum of -6. In order for the sum of all oxidation numbers to equal the overall charge (0), S must be +6. The oxidation numbers, then, are: <u>H: +1, S: +6, O: -2</u>

c. Rule #6 tells us that O is -2 here. Since the sum of all oxidation numbers must equal the overall charge (-2), C must be +4. The oxidation numbers, then, are: <u>C: +4, O: -2</u>

d. Rule #3 tells us that Mg is +2. This makes Cl -1. The oxidation numbers, then, are: <u>Mg: +2, Cl: -1</u>

e. Rule #2 tells us that K is +1. Rule #6 tells us that O is -2. This makes N +5. The oxidation numbers, then, are: <u>K: +1, N: +5, O: -2</u>

f. Rule #4 tells us that F is -1. That makes S +6. The oxidation numbers, then, are: <u>S: +6, F: -1</u>

g. We have to use the last resort here. Since Ir is not in groups 1-8 A, we cannot predict its charge in an ionic compound. However, we know that Cl is -1 in an ionic compound. This means Cl has an oxidation number of -1, making Ir +3. The oxidation numbers, then, are: <u>Ir: +3, Cl -1</u>

h. We cannot predict the charge of V, because it is not in groups 1-8 A. However, S is in group 6A, so the last resort rule says it will have a -2 oxidation number. Thus, <u>S is -2 and V is +2</u>.

3. To go from +3 to -1, you must gain negatives. This means the atom gained electrons, indicating that <u>it was reduced</u>. To go from +3 to -1, you must gain <u>4 electrons</u>.

4. To go from 0 to +2, you must lose negatives. This means the atom lost electrons, indicating that <u>it was oxidized</u>. To go from 0 to +2, you must lose <u>2 electrons</u>.

5. To go from -3 to 0, you must lose negatives. This means the atom lost electrons, indicating that <u>it was oxidized</u>. To go from -3 to 0, you must lose <u>3 electrons</u>.

6. To go from -1 to -3, you must gain negatives. This means the atom gained electrons, indicating that <u>it was reduced</u>. To go from -1 to -3, you must gain <u>2 electrons</u>.

7. a. <u>This is a redox reaction</u>, because Ni went from +2 to 0, indicating that <u>Ni was reduced</u>. At the same time, Cd went from 0 to +2, indicating that <u>Cd was oxidized</u>.

b. In this reaction, H stayed at +1, S stayed at +6, O stayed at -2, and N stayed at -3. Thus, <u>this is not a redox reaction</u>.

c. This is a redox reaction, because V went from +5 to +4, indicating that V was reduced. At the same time, Zn went from 0 to +2, indicating that Zn was oxidized.

d. In this reaction, Mg stayed at +2, O stayed at -2, and N stayed at +5, Na stayed at +1 and H stayed at +1. Thus, this is not a redox reaction.

e. This is a redox reaction, because Cl went from 0 to -1, indicating that Cl was reduced. At the same time, Na went from 0 to +1, indicating that Na was oxidized.

8. In this reaction, Cu^{2+} is going from an oxidation number of +2 to an oxidation number of 0. This indicates that it is gaining electrons. Thus, the solution holding aqueous Cu^{2+} will have electrons flowing into it. Electrons flow towards the Cu^{2+}, so that container is positive (it attracts electrons) and thus will be the cathode. The Co is going from an oxidation number of 0 to an oxidation number of +2. This means it loses electrons. Since it is losing electrons, the electrons are flowing away from the container holding the solid Co. This makes that container the negative side of the battery (it repels electrons), and it is thus the anode. The picture, then, looks like this:

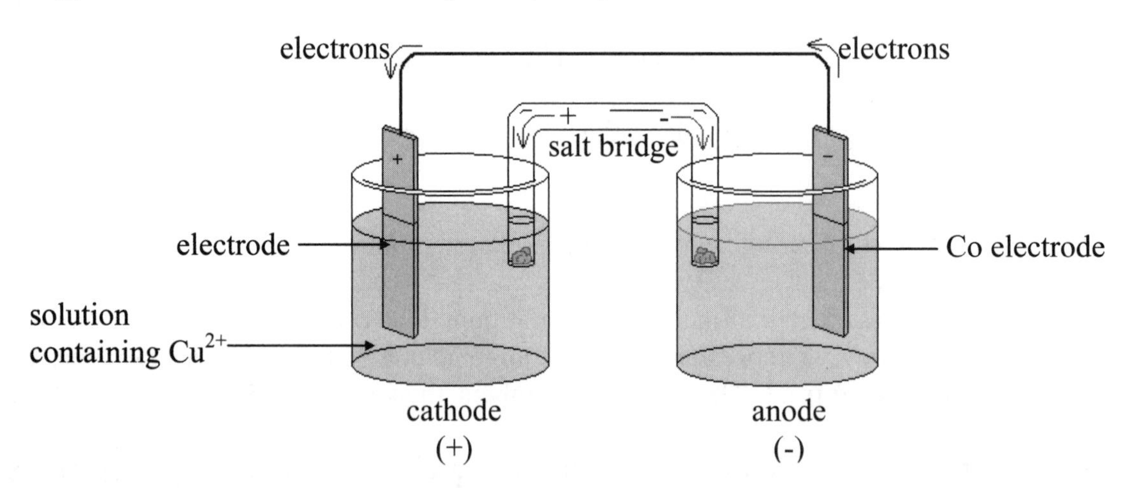

9. The first thing that we have to do is look at the reactants and see which is being oxidized and which is being reduced. Since aluminum goes from an oxidation number of 0 to an oxidation number of +3, it must be losing electrons, so it is oxidized. This means that the container that holds it will be the anode. The Fe^{3+}, on the other hand, gains electrons, because it goes from an oxidation number of +3 to an oxidation number of 0. This means that Fe^{3+} is reduced, and the container that holds it will be the cathode.

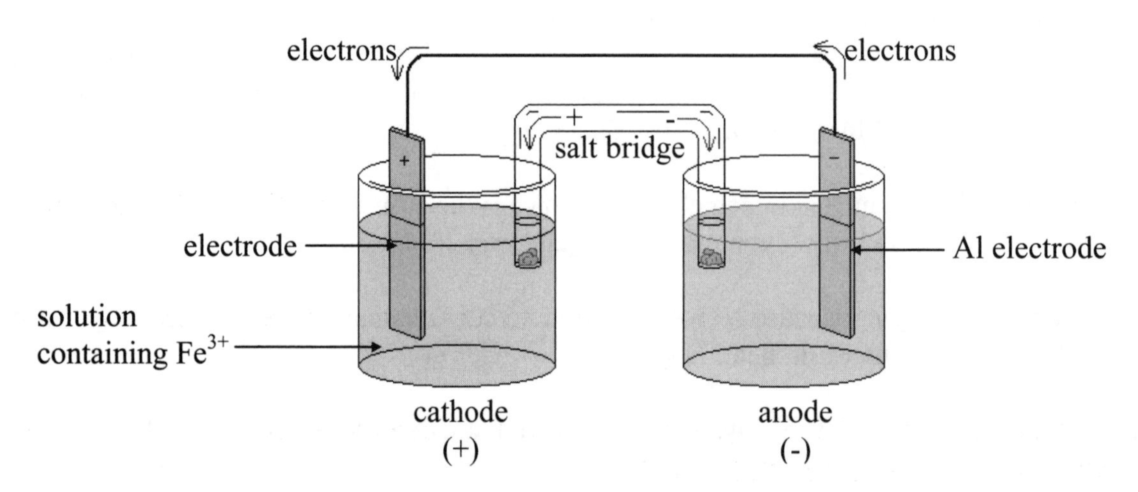

SOLUTIONS TO THE EXTRA PRACTICE PROBLEMS FOR MODULE #1

1. Remember, precision is determined by how many decimal places there are. The more decimal places, the more precise the number. Accuracy, on the other hand, tells us how close to the true value a measurement is. Thus, the first student is more accurate, while the second student is more precise.

2. $\frac{13.1\ \cancel{\text{g}}}{1} \times \frac{1\ \text{cg}}{0.01\ \cancel{\text{g}}} = \underline{1.31 \times 10^3\ \text{cg}}$

3. $\frac{45.1\ \cancel{\text{mL}}}{1} \times \frac{0.001\ \text{L}}{1\ \cancel{\text{mL}}} = \underline{0.0451\ \text{L}}$

4. To do this problem, you cannot directly convert miles to inches, because you have no direct relationship between the two. Thus, you must first convert to feet. Then you can convert to inches.

$$\frac{13.1\ \cancel{\text{miles}}}{1} \times \frac{5.280\text{x}10^3\ \text{feet}}{1\ \cancel{\text{mile}}} = 69{,}200\ \text{feet}$$

$$\frac{69{,}200\ \cancel{\text{feet}}}{1} \times \frac{12\ \text{inches}}{1\ \cancel{\text{foot}}} = \underline{8.30\text{x}10^5\ \text{inches}}$$

5. First, we need to plug the dimensions in the equation and get the volume:

$$V = (1.2\ \text{m}) \cdot (3.4\ \text{m}) \cdot (0.50\ \text{m}) = 2.0\ \text{m}^3$$

Although this **is** the volume of the sphere, it is not the answer, because the question asked for the volume in LITERS. Do we know of a way to convert from m^3 to liters? No, because we don't know of a relationship between them. We do know, however, that a mL is the same as a cm^3. Thus, we can convert from m^3 to cm^3, which then is the same as mL. Once we have mL, we can get to liters:

$$\frac{2.0\ \text{m}^3}{1} \times \left(\frac{1\ \text{cm}}{0.01\ \text{m}}\right)^3 = \frac{2.0\ \cancel{\text{m}^3}}{1} \times \frac{1\ \text{cm}^3}{0.000001\ \cancel{\text{m}^3}} = 2.0 \times 10^6\ \text{cm}^3 = 2.0 \times 10^6\ \text{mL}$$

$$\frac{2.0 \times 10^6\ \cancel{\text{mL}}}{1} \times \frac{0.001\ \text{liter}}{1\ \cancel{\text{mL}}} = \underline{2.0 \times 10^3\ \text{liters}}$$

6. The decimal point must be moved over three places to get it tot he right of the "3." Since the number is small, the exponent is negative. Thus, the answer is $\underline{3.41 \text{ x } 10^{-3}}$.

7. The exponent is positive, so this is a big number. The exponent also tells us to move the decimal point over 7 places. Thus, the answer is 11,800,000.

8. The two zeros are not significant. They are at the end of the number but not to the right of the decimal. Thus, there are two significant figures.

9. The first two zeros are not significant, but the one in between the significant figures is. Thus, there are three significant figures.

10. The last two zeros are significant, because they are at the end of the number and to the right of the decimal. The first two zeros are not significant. Thus, there are four significant figures.

11. This is an application of Equation (1.1):

$$\rho = \frac{m}{V}$$

$$\rho = \frac{23.13}{35.0\ mL} = \underline{0.661\ \frac{g}{mL}}$$

12. To do this problem, we must rearrange Equation (1.1):

$$\rho = \frac{m}{V}$$

$$2.12\ \frac{g}{mL} = \frac{45\ g}{V}$$

$$V = \frac{45\ \cancel{g}}{2.12\ \frac{\cancel{g}}{mL}} = \underline{21\ mL}$$

13. This is another application of the density equation, but we must first get the units consistent. The volume is in L, while the density is in g/mL. Thus, I need to convert L into mL:

$$\frac{0.67\ \cancel{L}}{1} \times \frac{1\ mL}{0.001\ \cancel{L}} = 670\ mL$$

Now that all units agree, we can use the equation for density to solve for mass.

$$\rho = \frac{m}{V}$$

$$19.3\ \frac{g}{mL} = \frac{m}{670\ mL}$$

$$m = (19.3\ \frac{g}{\cancel{mL}}) \cdot (670\ \cancel{mL}) = \underline{13{,}000\ g}$$

SOLUTIONS TO THE EXTRA PRACTICE PROBLEMS FOR MODULE #2

1. To solve this one, we simply need to use Equation (2.1):

$$^{\circ}C = \frac{5}{9}(^{\circ}F - 32)$$

$$^{\circ}C = \frac{5}{9}(34.5 - 32) = 1.4$$

Remember, 32 is exact. However, the precision of 34.5 limits the subtraction to the tenths place, which then leaves only two significant figures. Thus, 34.5 °F is the same as <u>1.4 °C</u>.

2. To solve this one, we start with Equation (2.1):

$$^{\circ}C = \frac{5}{9}(^{\circ}F - 32)$$

$$^{\circ}C = \frac{5}{9}(250.0 - 32) = 121.1\ ^{\circ}C$$

Now that we have the temperature in Celsius, we can convert to Kelvin:

$$K = 121.1\ ^{\circ}C + 273.15 = \underline{394.3\ K}$$

Neither of these numbers are exact, so we must use the rule of addition. The least precise number (121.1) has its last significant figure in the tenths place, so the answer must be reported to the tenths place.

3. To solve this problem, we must rearrange Equation (2.3). Remember, since energy is *removed*, the "q" of the liquid is *negative*:

$$\Delta T = \frac{q}{m \cdot c} = \frac{-1{,}625\ \text{J}}{(75\ \text{g}) \cdot (2.4\ \frac{\text{J}}{\text{g} \cdot {^{\circ}C}})} = -9.0\ ^{\circ}C$$

This is not the answer, however. The question wants the final temperature. Thus, we must rearrange Equation (2.4):

$$T_{final} = T_{initial} + \Delta T = 25.0\ ^{\circ}C + -9.0\ ^{\circ}C = \underline{16.0\ ^{\circ}C}$$

4. To solve this problem, we must rearrange Equation (2.3). This time, "q" is positive because heat is added:

$$\Delta T = \frac{q}{m \cdot c} = \frac{506\ \text{J}}{(50.0\ \text{g}) \cdot (4.184\ \frac{\text{J}}{\text{g} \cdot {^{\circ}C}})} = 2.42\ ^{\circ}C$$

This is not the answer, however. The question wants the final temperature. Thus, we must rearrange Equation (2.4):

$$T_{final} = T_{initial} + \Delta T = 15.0\ ^{o}C + 2.42\ ^{o}C = \underline{17.4\ ^{o}C}$$

5. This is a direct application of Equation (2.3):

$$q = m \cdot c \cdot \Delta T$$

$$q = (15.0\ g) \cdot \left(1.91\ \frac{J}{g \cdot {}^{o}C}\right) \cdot (25.0\ ^{o}C - 15.0\ ^{o}C)$$

$$q = (15.0\ \cancel{g}) \cdot \left(1.91\ \frac{J}{\cancel{g} \cdot {}^{o}\cancel{C}}\right) \cdot (10.0\ ^{o}\cancel{C}) = \underline{287\ J}$$

6. We can ignore the calorimeter in this problem, so that makes it a bit easier. We have all of the information that we need to calculate q_{water}, so we might as well start there:

$$q_{water} = m \cdot c \cdot \Delta T$$

$$q_{water} = (120.0\ g) \cdot \left(4.184\ \frac{J}{g \cdot {}^{o}C}\right) \cdot (37.0\ ^{o}C - 25.0\ ^{o}C)$$

$$q_{water} = (120.0\ \cancel{g}) \cdot \left(4.184\ \frac{J}{\cancel{g} \cdot {}^{o}\cancel{C}}\right) \cdot (12.0\ ^{o}\cancel{C}) = 6{,}020\ J$$

Since we can assume $q_{calorimeter} = 0$, we can now determine q_{metal}:

$$-q_{metal} = q_{water} + q_{calorimeter} = 6{,}020\ J + 0$$

$$q_{metal} = -6{,}020\ J$$

The value for q_{metal} is negative because the metal *lost* energy. We can calculate the ΔT of the metal:

$$\Delta T_{metal} = T_{final} - T_{initial} = 37.0\ ^{o}C - 100.0\ ^{o}C = -63.0\ ^{o}C$$

We now have all the information we need to calculate the specific heat of the metal:

$$c = \frac{q}{m \cdot \Delta T}$$

$$c = \frac{-6{,}020\ J}{500.0\ g \cdot -63.0\ ^{o}C} = \underline{0.191\ \frac{J}{g \cdot {}^{o}C}}$$

7. We are given the details of the calorimeter in this problem and we are not told that we can ignore it. Thus, to solve this problem, we will need to determine the heat gained by the water *and* the heat gained by the calorimeter. Let's start with the calorimeter:

$$q_{\text{calorimeter}} = m \cdot c \cdot \Delta T$$

$$q_{\text{calorimeter}} = (5.0\ \text{g}) \cdot \left(1.40\ \frac{\text{J}}{\text{g} \cdot {}^{\circ}\text{C}}\right) \cdot (37.0\ {}^{\circ}\text{C} - 25.0\ {}^{\circ}\text{C})$$

$$q_{\text{calorimeter}} = (5.0\ \cancel{\text{g}}) \cdot \left(1.40\ \frac{\text{J}}{\cancel{\text{g}} \cdot {}^{\circ}\cancel{\text{C}}}\right) \cdot (12.0\ {}^{\circ}\cancel{\text{C}}) = 84\ \text{J}$$

Now we can move on to the water:

$$q_{\text{water}} = m \cdot c \cdot \Delta T$$

$$q_{\text{water}} = (150.0\ \cancel{\text{g}}) \cdot \left(4.184 \frac{\text{J}}{\cancel{\text{g}} \cdot {}^{\circ}\cancel{\text{C}}}\right) \cdot (12.0\ {}^{\circ}\cancel{\text{C}}) = 7{,}530\ \text{J}$$

Now that we have calculated the "q's" of the water and calorimeter, we can use Equation (2.5) to determine the "q" of the metal.

$$-q_{\text{metal}} = q_{\text{water}} + q_{\text{calorimeter}} = 7{,}530\ \text{J} + 84\ \text{J} = 7{,}610\ \text{J}$$

$$q_{\text{metal}} = -7{,}610\ \text{J}$$

Now that we have "q_{metal}," we can determine ΔT for the metal and then use Equation (2.3) to determine the specific heat:

$$\Delta T_{\text{metal}} = T_{\text{final}} - T_{\text{initial}} = 37.0\ {}^{\circ}\text{C} - 112.4\ {}^{\circ}\text{C} = -75.4\ {}^{\circ}\text{C}$$

Notice that this ΔT is negative, because the object cooled down.

$$c = \frac{q}{m \cdot \Delta T}$$

$$c = \frac{-7{,}610\ \text{J}}{250.0\,\text{g} \cdot (-75.4\ {}^{\circ}\text{C})} = \underline{0.404\ \frac{\text{J}}{\text{g} \cdot {}^{\circ}\text{C}}}$$

8. When we look at this problem, we see that we're trying to discover the specific heat of the liquid in the calorimeter, not the specific heat of the metal, as is usually the case. Instead, the specific heat of the metal is given. In order to find the liquid's specific heat, we need to find out "q_{liquid}". How can we do that? Well, we have enough information to calculate "q_{metal}" and "$q_{\text{calorimeter}}$", so we can start there.

$$q_{metal} = m \cdot c \cdot \Delta T$$

$$q_{metal} = (50.0\text{ g}) \cdot \left(0.506 \frac{\text{J}}{\text{g} \cdot {}^\circ\text{C}}\right) \cdot (20.0\ {}^\circ\text{C} - 60.0\ {}^\circ\text{C})$$

$$q_{metal} = (50.0\text{ g}) \cdot \left(0.506 \frac{\text{J}}{\text{g} \cdot {}^\circ\text{€}}\right) \cdot (-40.0\ {}^\circ\text{€}) = -1{,}010\text{ J}$$

Notice that the heat is negative. You should expect that because the metal lost energy. We can now determine the heat gained by the calorimeter:

$$q_{calorimeter} = m \cdot c \cdot \Delta T$$

$$q_{calorimeter} = (5.0\text{ g}) \cdot \left(1.40 \frac{\text{J}}{\text{g} \cdot {}^\circ\text{€}}\right) \cdot (20.0\ {}^\circ\text{€} - 15.0\ {}^\circ\text{€}) = 35\text{ J}$$

Now we can use Equation (2.5) to determine "q_{liquid}". Since the unknown liquid is replacing water in the calorimeter, we can replace "q_{water}" with "q_{liquid}" in the equation:

$$-q_{metal} = q_{liquid} + q_{calorimeter}$$

$$q_{liquid} = -q_{metal} - q_{calorimeter} = -(-1{,}010\text{ J}) - 35\text{ J} = 980\text{ J}$$

Since the "q" of the metal has its last significant figure in the tens place, the answer must be reported to the tens place. That's why 975 was rounded up to 980. We can now determine the specific heat of the liquid:

$$c = \frac{q}{m \cdot \Delta T}$$

$$c = \frac{980\text{ J}}{(100.0\text{ g}) \cdot (5.0\ {}^\circ\text{C})} = \underline{2.0 \frac{\text{J}}{\text{g} \cdot {}^\circ\text{C}}}$$

9. In this problem, we're trying to discover the specific heat of the calorimeter. In order to find the calorimeter's specific heat, we need to find out "$q_{calorimeter}$". How can we do that? Well, we have enough information to calculate "q_{metal}" and "q_{liquid}", so we can start there.

$$q_{\text{metal}} = m \cdot c \cdot \Delta T$$

$$q_{\text{metal}} = (150.0\ \text{g}) \cdot \left(0.560 \frac{\text{J}}{\text{g} \cdot {}^{\circ}\text{C}}\right) \cdot (29.0\ {}^{\circ}\text{C} - 55.0\ {}^{\circ}\text{C})$$

$$q_{\text{metal}} = (150.0\ \text{g}) \cdot \left(0.560 \frac{\text{J}}{\text{g} \cdot {}^{\circ}\text{€}}\right) \cdot (-26.0\ {}^{\circ}\text{€}) = -2{,}180\ \text{J}$$

We can now determine the heat gained by the water:

$$q_{\text{water}} = m \cdot c \cdot \Delta T$$

$$q_{\text{water}} = (40.0\ \text{g}) \cdot \left(4.184 \frac{\text{J}}{\text{g} \cdot {}^{\circ}\text{€}}\right) \cdot (29.0\ {}^{\circ}\text{€} - 25.0\ {}^{\circ}\text{€}) = 670\ \text{J}$$

Now we can use Equation (2.5) to determine "$q_{\text{calorimeter}}$".

$$-q_{\text{metal}} = q_{\text{water}} + q_{\text{calorimeter}}$$

$$q_{\text{calorimeter}} = -q_{\text{metal}} - q_{\text{water}} = -(-2{,}180\ \text{J}) - 670\ \text{J} = 1{,}510\ \text{J}$$

Since both values for "q" have their last significant figure in the tens place, the answer must be reported to the tens place. Now we can get the specific heat of the calorimeter:

$$c = \frac{q}{m \cdot \Delta T}$$

$$c = \frac{1{,}510\ \text{J}}{(5.0\ \text{g}) \cdot (4.0\ {}^{\circ}\text{C})} = \underline{76 \frac{\text{J}}{\text{g} \cdot {}^{\circ}\text{C}}}$$

10. In this case, we are trying to determine the mass of water in the calorimeter. Thus, we will need to determine "q_{water}." To do that, we need to get "q_{metal}" and "$q_{\text{calorimeter}}$."

$$q_{\text{metal}} = m \cdot c \cdot \Delta T$$

$$q_{\text{metal}} = (75.0\ \text{g}) \cdot \left(0.3851 \frac{\text{J}}{\text{g} \cdot {}^{\circ}\text{C}}\right) \cdot (27.5\ {}^{\circ}\text{C} - 100.0\ {}^{\circ}\text{C})$$

$$q_{\text{metal}} = (75.0\ \text{g}) \cdot \left(0.3851\ \frac{\text{J}}{\text{g} \cdot {}^{\circ}\text{C}}\right) \cdot (-72.5\ {}^{\circ}\text{C}) = -2{,}090\ \text{J}$$

$$q_{\text{calorimeter}} = m \cdot c \cdot \Delta T$$

$$q_{\text{calorimeter}} = (5.0\ \text{g}) \cdot \left(1.40\ \frac{\text{J}}{\text{g} \cdot {}^{\circ}\text{C}}\right) \cdot (27.5\ {}^{\circ}\text{C} - 25.0\ {}^{\circ}\text{C}) = 18\ \text{J}$$

Now we can use the calorimetry equation to get "q_{water}":

$$-q_{\text{metal}} = q_{\text{water}} + q_{\text{calorimeter}}$$

$$q_{\text{water}} = -q_{\text{metal}} - q_{\text{calorimeter}} = -(-2{,}090\ \text{J}) - 18\ \text{J} = 2{,}070\ \text{J}$$

We can get the mass of water by rearranging Equation (2.3):

$$m = \frac{q}{c \cdot \Delta T}$$

$$m = \frac{2{,}070\ \text{J}}{\left(4.184\ \frac{\text{J}}{\text{g} \cdot {}^{\circ}\text{C}}\right) \cdot (2.5\ {}^{\circ}\text{C})} = \underline{2.0 \times 10^{2}\ \text{g}}$$

Since the ΔT restricts the number of significant figures to two, the only way to properly report the answer is with scientific notation.

SOLUTIONS TO THE EXTRA PRACTICE PROBLEMS FOR MODULE #3

1. A compound is ionic if, when dissolved in water, it conducts electricity. If it does not conduct electricity, the compound is covalent. Thus, you can perform experiments like Experiment 3.2 to determine whether a compound conducts electricity when dissolved in pure water. That will determine whether it is ionic or covalent.

2. If a compound has a metal in it, it must be ionic. If it has no metals, it is covalent.

3. All elements that lie to the left of the jagged line on the chart are metals, while all elements to the right of the jagged line are nonmetals. A molecule is covalent only if it has no metals in it. Thus, SO_2 and C_2H_4O are covalent.

4. All elements that lie to the left of the jagged line on the chart are metals, while all elements to the right of the jagged line are nonmetals. A molecule is ionic only if it has metals in it. Thus, $RbNO_3$ and Na_2SO_4 are ionic.

5. The reaction starts with 100.0 g + 100.0 g = 200.0 g of matter; thus, there must be 200.0 g of matter after everything is finished. According to the problem, these amounts of calcium and chlorine made 156.5 g of calcium chloride along with left over calcium. Since all 200.0 g must be accounted for, the remaining mass must be in the calcium:

$$\text{Mass of calcium} = \text{Total mass} - \text{Mass of product}$$

$$\text{Mass of calcium} = 200.0\text{ g} - 156.5\text{ g} = 43.5\text{ g}$$

By the law of mass conservation, then, there were 43.5 g of calcium left over. Since we started out with 100.0 g of calcium and there were 43.5 g left over, then only 100.0 g - 43.5 g = 56.5 g were actually used to make calcium chloride. Thus, the proper recipe for making 156.5 g of calcium chloride is to add 100.0 g of chlorine to 56.5 g of calcium.

The problem, however, asks us the recipe for making 1.000 kg, or 1.000×10^3 g of calcium chloride. Therefore, we need to determine how much to increase the amount of ingredients in order to make this larger amount:

$$156.5\text{g} \cdot \text{x} = 1.000 \times 10^3\text{ g}$$

$$\text{x} = \frac{1.000 \times 10^3\text{ g}}{156.5\text{g}} = 6.390$$

To make 1.0 kg, then, we just multiply the amount of each component by 6.390:

$$\text{Mass of chlorine} = 100.0\text{ g} \cdot 6.390 = 639.0\text{ g}$$

$$\text{Mass of calcium} = 56.5\text{ g} \cdot 6.390 = 361\text{ g}$$

You need 639.0 g of chlorine and 361 g of calcium to make 1.000 kg of calcium chloride.

6. The reaction starts with 50.0 g + 50.0 g = 100.0 g of matter; thus, there must be 100.0 g of matter after everything is finished. According to the problem, these amounts of nitrogen and hydrogen made 60.7 g of ammonia along with left over hydrogen. Since all 100.0 g must be accounted for, the remaining mass must be in the hydrogen:

$$\text{Mass of hydrogen} = \text{Total mass} - \text{Mass of product}$$

$$\text{Mass of hydrogen} = 100.0 \text{ g} - 60.7 \text{ g} = 39.3 \text{ g}$$

By the law of mass conservation, then, there were 39.3 g of hydrogen left over. Since we started out with 50.0 g of hydrogen and there were 39.3 g left over, then only 50.0 g - 39.3 g = 10.7 g were actually used to make ammonia. Thus, the proper recipe for making 60.7 g of ammonia is to add 50.0 g of nitrogen to 10.7 g of hydrogen.

The problem, however, asks us the recipe for making 100.0 g of ammonia. Therefore, we need to determine how much to increase the amount of ingredients in order to make this larger amount:

$$60.7 \text{ g} \cdot \text{x} = 100.0 \text{ g}$$

$$\text{x} = \frac{100.0 \text{ g}}{60.7 \text{ g}} = 1.65$$

To make 100.0 g, then, we just multiply the amount of each component by 1.65:

$$\text{Mass of nitrogen} = 50.0 \text{ g} \cdot 1.65 = 82.5 \text{ g}$$

$$\text{Mass of hydrogen} = 10.7 \text{ g} \cdot 1.65 = 17.7 \text{ g}$$

You need 82.5 g of nitrogen and 17.7 g of hydrogen to make 100.0 g of ammonia. The numbers actually add to 100.2 because of the rounding done for significant figures.

7. In this problem, we already have the recipe for making methane with no leftovers (12.0 g carbon + 4.0 g hydrogen). When the chemist adds 50.0 grams of carbon, he is increasing the amount by:

$$12.0 \text{ g} \cdot \text{x} = 50.0 \text{ g}$$

$$\text{x} = \frac{50.0 \text{ g}}{12.0 \text{ g}} = 4.17$$

If the chemist added 4.17 times as much carbon, he should also add 4.17 times as much hydrogen:

$$4.0 \text{ g} \cdot 4.17 = 17 \text{ g}$$

However, the chemist added 50.0 g of hydrogen. This means that 50.0 g - 17 g = 33 g of hydrogen will not react. Thus, 33 g of hydrogen will be left over once the methane is made.

8. The reaction starts with 54.0 g + 100.0 g = 154.0 g of matter; thus, there must be 154.0 g of matter after everything is finished. According to the problem, these amounts of aluminum and sulfur made 150.3 g of aluminum sulfide along with left over sulfur. Since all 154.0 g must be accounted for, the remaining mass must be in the sulfur:

$$\text{Mass of sulfur} = \text{Total mass} - \text{Mass of product}$$

$$\text{Mass of sulfur} = 154.0 \text{ g} - 150.3 \text{ g} = 3.7 \text{ g}$$

By the law of mass conservation, then, there were 3.7 g of sulfur left over. Since we started out with 100.0 g of sulfur and there were 3.7 g left over, then only 100.0 g - 3.7 g = 96.3 g were actually used to make aluminum sulfide. Thus, the proper recipe for making 150.3 g of aluminum sulfide is to add 54.0 g of aluminum to 96.3 g of sulfur.

The problem, however, asks us the recipe for making 100.0 g of aluminum sulfide. Therefore, we need to determine how much to decrease the amount of ingredients in order to make this larger amount:

$$150.3 \text{g} \cdot \text{x} = 100.0 \text{ g}$$

$$\text{x} = \frac{100.0 \text{ g}}{150.3 \text{ g}} = 0.6653$$

To make 100.0 g, then, we just multiply the amount of each component by 0.6653:

$$\text{Mass of aluminum} = 54.0 \text{g} \cdot 0.6653 = 35.9 \text{ g}$$

$$\text{Mass of sulfur} = 96.3 \text{ g} \cdot 0.6653 = 64.1 \text{ g}$$

You need 35.9 g of aluminum and 64.1 g of sulfur to make 100.0 g of aluminum sulfide.

9. a. P_2O_6 b. N_2H_4 c. NO

10. a. This is a covalent compound, so prefixes must be used. The name is sulfur dioxide.

b. This is an ionic compound, so no prefixes are needed. The name is calcium chloride.

c. This is a covalent compound, so prefixes are needed. The name is oxygen dichloride.

d. This is an ionic compound, so no prefixes are needed. The name is sodium sulfide.

SOLUTIONS TO THE EXTRA PRACTICE PROBLEMS FOR MODULE #4

1. The liquid turned into a gas. If it had turned into a solid, the molecules would have slowed down and moved closer together. Even if the liquid was water, the molecules would have still slowed down.

2. a. The amount of cereal and milk is different in different parts of the bowl, so this is a heterogeneous mixture.

b. This is a compound, as it is represented by a single chemical formula that contains more than one type of atom.

c. This is an element, as it can be found on the periodic chart (Cu).

d. This is a homogeneous mixture, as the amounts of Kool-Aid® and water are the same throughout.

3. a. This is a physical change, as the glass is all still glass.

b. This is a chemical change. It would be pretty hard to "unbake" bread.

c. This is a physical change, as all phase changes are physical.

d. This is a chemical change. It would be pretty hard to "unburn" gasoline.

4. Chlorine is a homonuclear diatomic. Thus, it is Cl_2.

5. $2HCl\ (aq) + Ca\ (s) \rightarrow CaCl_2\ (aq) + H_2\ (g)$

6. $2HBr\ (aq) + 2Na\ (s) \rightarrow 2NaBr\ (s) + H_2\ (g)$

7. $2AgCl\ (aq) + Zn\ (s) \rightarrow ZnCl_2\ (aq) + 2Ag\ (s)$

8. $MgCl_2\ (aq) + H_2S\ (aq) \rightarrow MgS\ (s) + 2HCl\ (aq)$

9. $C_5H_{12} + 8O_2 \rightarrow 5CO_2 + 6H_2O$

10. $18CO_2\ (g) + 16H_2O\ (l) \rightarrow C_{18}H_{32}O_{16}\ (s) + 18O_2\ (g)$

SOLUTIONS TO THE EXTRA PRACTICE PROBLEMS FOR MODULE #5

1. $\underline{2FeCl_3 + 3Na_2CO_3 \rightarrow Fe_2C_3O_9 + 6NaCl}$

2. $\underline{2NH_4Cl + BaO_2H_2 \rightarrow BaCl_2 + 2NH_3 + 2H_2O}$

3. $\underline{P_4O_{10} + 6H_2O \rightarrow 4H_3PO_4}$

4. $\underline{4CoS_2 + 11O_2 \rightarrow 2Co_2O_3 + 8SO_2}$

5. Since H and O are homonuclear diatomics, we must have H_2 and O_2 as reactants:

$$C + H_2 + O_2 \rightarrow C_2H_4O$$

Now we balance:

$$\underline{4C + 4H_2 + O_2 \rightarrow 2C_2H_4O}$$

6. The chemical equation tells us that

$$4 \text{ moles C} = 2 \text{ moles } C_2H_4O$$

That allows us to convert from C to C_2H_4O:

$$\frac{15 \cancel{\text{moles } C_2H_4O}}{1} \times \frac{4 \text{ moles C}}{2 \cancel{\text{moles } C_2H_4O}} = \underline{3.0 \times 10^1 \text{ moles C}}$$

7. We will assume that this is complete combustion, which involved adding oxygen and producing carbon dioxide and water:

$$C_{16}H_{34} + O_2 \rightarrow CO_2 + H_2O$$

Now we have to balance:

$$\underline{2C_{16}H_{34} + 49O_2 \rightarrow 32CO_2 + 34H_2O}$$

8. To do this, we first must come up with a balanced chemical equation. We will assume that this is complete combustion, which involved adding oxygen and producing carbon dioxide and water:

$$C_{16}H_{24} + O_2 \rightarrow CO_2 + H_2O$$

Now we have to balance:

$$C_{16}H_{24} + 22O_2 \rightarrow 16CO_2 + 12H_2O$$

This tells us:

$$1 \text{ mole } C_{16}H_{24} = 16 \text{ moles } CO_2$$

This allows us to convert from moles of $C_{16}H_{24}$ to moles of CO_2:

$$\frac{13.2 \cancel{\text{moles } C_{16}H_{24}}}{1} \times \frac{16 \text{ moles } CO_2}{1 \cancel{\text{mole } C_{16}H_{24}}} = \underline{211 \text{ moles } CO_2}$$

9. This molecule has 2 K's (39.1 amu), 1 C (12.0 amu), and 3 O's (16.0 amu). The molecular mass, then, is:

$$\text{Mass of } K_2CO_3 = 2 \text{ x } 39.1 \text{ amu} + 12.0 \text{ amu} + 3 \text{ x } 16.0 \text{ amu} = 138.2 \text{ amu}$$

Converting to grams:

$$\frac{138.2 \cancel{\text{amu}}}{1} \times \frac{1.66 \times 10^{-24} \text{ g}}{1.00 \cancel{\text{amu}}} = 2.29 \times 10^{-22} \text{ g}$$

The mass is $\underline{2.29 \text{ x } 10^{-22} \text{ g}}$.

10. This molecule has 2 H's (1.01 amu), 1 C (12.0 amu), and 3 O's (16.0 amu). The molecular mass, then, is:

$$\text{Mass of } H_2CO_3 = 2 \text{ x } 1.01 \text{ amu} + 12.0 \text{ amu} + 3 \text{ x } 16.0 \text{ amu} = 62.0 \text{ amu}$$

This tells us that:

$$62.0 \text{ grams } H_2CO_3 = 1 \text{ mole } H_2CO_3$$

This is the conversion relationship we need for converting grams into moles:

$$\frac{125 \cancel{\text{g } H_2CO_3}}{1} \times \frac{1 \text{ mole } H_2CO_3}{62.0 \cancel{\text{g } H_2CO_3}} = \underline{2.02 \text{ moles } H_2CO_3}$$

11. This is just another conversion problem, but this time we are converting moles into grams. We still need to determine the molecular mass of $CaCl_2$ first, however:

$$\text{Mass of } CaCl_2 = 1 \text{ x } 40.1 \text{ amu} + 2 \text{ x } 35.5 \text{ amu} = 111.1 \text{ amu}$$

This means:

$$111.1 \text{ grams } CaCl_2 = 1 \text{ mole } CaCl_2$$

Now we can do our conversion:

$$\frac{0.172 \cancel{\text{moles } CaCl_2}}{1} \times \frac{111.1 \text{ g } CaCl_2}{1 \cancel{\text{mole } CaCl_2}} = \underline{19.1 \text{ g } CaCl_2}$$

12. We start by determining the mass of a molecule:

$$\text{Mass of KCl} = 1 \text{ x } 39.1 \text{ amu} + 1 \text{ x } 35.5 \text{ amu} = 74.6 \text{ amu}$$

This tells us that:

$$74.6 \text{ grams KCl} = 1 \text{ mole KCl}$$

This is the conversion relationship we need for converting grams into moles:

$$\frac{50.0 \cancel{\text{g KCl}}}{1} \times \frac{1 \text{ mole KCl}}{74.6 \cancel{\text{g KCl}}} = \underline{0.670 \text{ moles KCl}}$$

13. Once again, we start by determining the mass of a molecule:

$$\text{Mass of } CO_2 = 1 \text{ x } 12.0 \text{ amu} + 2 \text{ x } 16.0 \text{ amu} = 44.0 \text{ amu}$$

This means:

$$44.0 \text{ grams } CO_2 = 1 \text{ mole } CO_2$$

Now we can do our conversion:

$$\frac{4.51 \cancel{\text{moles } CO_2}}{1} \times \frac{44.0 \text{ g } CO_2}{1 \cancel{\text{mole } CO_2}} = 198 \text{ g } CO_2$$

SOLUTIONS TO THE EXTRA PRACTICE PROBLEMS FOR MODULE #6

1. This problem asks us to determine the mass of one substance in the equation given the mass of another substance. Thus, we must use stoichiometry. The first thing that we must do, then, is convert to moles:

Mass of HNO_3 = 1 x 1.01 amu + 1 x 14.0 amu + 3 x 16.0 amu = 63.0 amu

1 mole of HNO_3 = 63.0 g HNO_3

$$\frac{5.89\times10^{6}\ \cancel{\text{g}\,HNO_3}}{1}\times\frac{1\text{ mole }HNO_3}{63.0\ \cancel{\text{g}\,HNO_3}} = 9.35\times10^{4}\text{ moles }HNO_3$$

The chemical equation tells us:

$$3\text{ moles }NO_2 = 2\text{ moles }HNO_3$$

Thus, we can use it as a conversion relationship to convert moles of HNO_3 into moles of NO_2:

$$\frac{9.35\times10^{4}\ \cancel{\text{moles}\,HNO_3}}{1}\times\frac{3\text{ moles }NO_2}{2\ \cancel{\text{moles }HNO_3}} = 1.40\times10^{5}\text{ moles }NO_2$$

Now we can convert to grams of NO_2:

Mass of NO_2 = 1 x 14.0 amu + 2 x 16.0 amu = 46.0 amu

1 mole NO_2 = 46.0 g NO_2

$$\frac{1.40\times10^{5}\ \cancel{\text{moles }NO_2}}{1}\times\frac{46.0\text{ g }NO_2}{1\ \cancel{\text{mole }NO_2}} = 6.44\times10^{6}\text{ g }NO_2$$

2. The first thing that we must do is convert to moles:

Mass of H_2 = 2 x 1.01 amu = 2.02 amu

1 mole of H_2 = 2.02 g H_2

$$\frac{1.0\times10^{6}\ \cancel{\text{g }H_2}}{1}\times\frac{1\text{ mole }H_2}{2.02\ \cancel{\text{g }H_2}} = 5.0\times10^{5}\text{ moles }H_2$$

The chemical equation tells us :

$$3\text{ moles }H_2 = 1\text{ mole W}$$

Thus, we can use it as a conversion relationship to convert moles of H_2 into moles of W:

$$\frac{5.0\times10^{5}\ \cancel{\text{moles}\,H_2}}{1}\times\frac{1\text{ mole W}}{3\ \cancel{\text{moles }H_2}} = 1.7\times10^{5}\text{ moles W}$$

Now we can convert to grams W:

$$\frac{1.7 \times 10^5 \ \cancel{\text{moles W}}}{1} \times \frac{183.9 \text{ g W}}{1 \ \cancel{\text{mole W}}} = \underline{3.1 \times 10^7 \text{ g W}}$$

3. We already have moles of H_2, so we can use that to convert to moles of WO_3 using the chemical equation:

$$3 \text{ moles } H_2 = 1 \text{ mole } WO_3$$

$$\frac{5.0 \times 10^5 \ \cancel{\text{moles} H_2}}{1} \times \frac{1 \text{ mole } WO_3}{3 \ \cancel{\text{moles } H_2}} = 1.7 \times 10^5 \text{ moles } WO_3$$

Now we can convert to grams WO_3:

Mass of WO_3 = 1 x 183.9 amu + 3 x 16.0 amu = 231.9 amu

1 mole WO_3 = 231.9 g WO_3

$$\frac{1.7 \times 10^5 \ \cancel{\text{moles } WO_3}}{1} \times \frac{231.9 \text{ g } WO_3}{1 \ \cancel{\text{mole } WO_3}} = \underline{3.9 \times 10^7 \text{ g } WO_3}$$

4. The first thing that we must do is convert to moles:

$$\frac{16{,}000 \ \cancel{\text{g } CCl_4}}{1} \times \frac{1 \text{ mole } CCl_4}{154 \ \cancel{\text{g } CCl_4}} = 1.0 \times 10^2 \text{ moles } CCl_4$$

The chemical equation tells us:

$$1 \text{ mole } CS_2 = 1 \text{ mole } CCl_4$$

Thus, we can use it as a conversion relationship to convert moles of CCl_4 into moles of CS_2:

$$\frac{1.0 \times 10^2 \ \cancel{\text{moles} CCl_4}}{1} \times \frac{1 \text{ mole } CS_2}{1 \ \cancel{\text{mole } CCl_4}} = 1.0 \times 10^2 \text{ moles } CS_2$$

Now we can convert to grams CS_2:

$$\frac{1.0 \times 10^2 \ \cancel{\text{moles } CS_2}}{1} \times \frac{76.2 \text{ g } CS_2}{1 \ \cancel{\text{mole } CS_2}} = \underline{7{,}600 \text{ g } CS_2}$$

5. The first thing that we must do is convert to moles:

$$\frac{2.50 \times 10^3 \ \cancel{\text{g } Ca_3P_2O_8}}{1} \times \frac{1 \text{ mole } Ca_3P_2O_8}{310.3 \ \cancel{\text{g } Ca_3P_2O_8}} = 8.06 \text{ moles } Ca_3P_2O_8$$

The chemical equation tells us:

$$1 \text{ mole } Ca_3P_2O_8 = 1 \text{ mole } CaH_4P_2O_8$$

Thus, we can use it as a conversion relationship to convert moles of $Ca_3P_2O_8$ into moles of $CaH_4P_2O_8$:

$$\frac{8.06 \cancel{\text{moles } Ca_3P_2O_8}}{1} \times \frac{1 \text{ mole } CaH_4P_2O_8}{1 \cancel{\text{mole } Ca_3P_2O_8}} = 8.06 \text{ moles } CaH_4P_2O_8$$

Now we can convert to grams $CaH_4P_2O_8$:

$$\frac{8.06 \cancel{\text{moles } CaH_4P_2O_8}}{1} \times \frac{234.1 \text{ g } CaH_4P_2O_8}{1 \cancel{\text{mole } CaH_4P_2O_8}} = \underline{1{,}890 \text{ g } CaH_4P_2O_8}$$

6. The first thing that we must do is convert to moles:

$$\frac{5.6 \times 10^4 \cancel{\text{ g } Mg(OH)_2}}{1} \times \frac{1 \text{ mole } Mg(OH)_2}{58.3 \cancel{\text{ g } Mg(OH)_2}} = 960 \text{ moles } Mg(OH)_2$$

The chemical equation tells us:

$$1 \text{ mole } Mg(OH)_2 = 1 \text{ mole } Mg$$

Thus, we can use it as a conversion relationship to convert moles of $Mg(OH)_2$ into moles of Mg:

$$\frac{960 \cancel{\text{ moles } Mg(OH)_2}}{1} \times \frac{1 \text{ mole } Mg}{1 \cancel{\text{ mole } Mg(OH)_2}} = 960 \text{ moles } Mg$$

Now we can convert to grams Mg:

$$\frac{960 \cancel{\text{ moles } Mg}}{1} \times \frac{24.3 \text{ g } Mg}{1 \cancel{\text{ mole } Mg}} = \underline{23{,}000 \text{ g } Mg}$$

7. Empirical formulas have no common factors in the subscripts. Thus, $\underline{SO_3 \text{ and } K_2S_2O_3}$ are empirical formulas.

8. We can determine the amount of oxygen from the Law of Mass Conservation. Since we started with 100.0 g of matter, we must end with 100.0 g of matter. Thus, there were 100.0 g - 39.7 g - 27.9 g = 32.4 g of oxygen. Since there is potassium, manganese, and oxygen in the sample, the decomposition reaction is:

$$K_xMn_yO_z \rightarrow K + Mn + O_2$$

In order to get the stoichiometric coefficients, we determine the number of moles produced in the decomposition reaction.

$$\frac{39.7\ \cancel{g\ K}}{1} \times \frac{1\ \text{mole K}}{39.1\ \cancel{g\ K}} = 1.02\ \text{moles K}$$

$$\frac{27.9\ \cancel{g\ Mn}}{1} \times \frac{1\ \text{mole Mn}}{54.9\ \cancel{g\ Mn}} = 0.508\ \text{moles Mn}$$

$$\frac{32.4\ \cancel{g\ O_2}}{1} \times \frac{1\ \text{mole } O_2}{32.0\ \cancel{g\ O_2}} = 1.01\ \text{moles } O_2$$

So the equation looks like this:

$$K_xMn_yO_z \rightarrow 1.02K + 0.508Mn + 1.01O_2$$

The problem with this equation is that the coefficients are not integers. We need to turn them into integers in order to be able to determine the empirical formula of the reactant. How do we do that? We divide by the smallest number:

$$K_xMn_yO_z \rightarrow \frac{1.02}{0.508}K + \frac{0.508}{0.508}Mn + \frac{1.01}{0.508}O_2$$

$$K_xMn_yO_z \rightarrow 2.01K + 1.00Mn + 1.99O_2$$

We can since 2.01 is very, very close to 2 and 1.99 is very, very close to 2, we can probably assume that they should both really be 2. Thus, the formula that balances the equation, $\underline{K_2MnO_4}$. This is already an empirical formula.

9. Since the mass of the molecule is the same as the mass of the empirical formula, the empirical formula is also the molecular formula: $\underline{K_2MnO_4}$.

10. The mass of CH_2O is 30.0 amu. To reach the mass of the molecule, I will need to multiply each element by 2. Thus, the molecular formula is $\underline{C_2H_4O_2}$.

SOLUTIONS TO THE EXTRA PRACTICE PROBLEMS FOR MODULE #7

1. 2. a. Looking at the chart, Na has an atomic number of 11. This means it has 11 protons and 11 electrons. Its mass number, according to the problem, is 23. If it has 23 total protons + neutrons and it has 11 protons, then it has 23 - 11 = 12 neutrons.

b. Looking at the chart, Ar has an atomic number of 18. This means it has 18 protons and 18 electrons. Its mass number, according to the problem, is 40. If it has 40 total protons + neutrons and it has 18 protons, then it has 40 - 18 = 22 neutrons.

c. Looking at the chart, Fe has an atomic number of 26. This means it has 26 protons and 26 electrons. Its mass number, according to the problem, is 55. If it has 55 total protons + neutrons and it has 26 protons, then it has 55 - 26 = 29 neutrons.

d. Looking at the chart, U has an atomic number of 92. This means it has 92 protons and 92 electrons. Its mass number, according to the problem, is 238. If it has 238 total protons + neutrons and it has 92 protons, then it has 238 - 92 = 146 neutrons.

2. Remember ROY G. BIV. This is the order of visible light wavelengths from the largest to the smallest. Thus, the yellow light bulb has larger wavelengths. When wavelength is large, however, frequency is small; thus, the blue light has the highest frequency. The higher the frequency, the higher the energy, so the blue light also has the highest energy.

3. We are given the fact that 1 nm = 10^{-9} m. Thus, we can first convert to meters:

$$\frac{425\ \cancel{\text{nm}}}{1} \times \frac{10^{-9}\ \text{m}}{1\ \cancel{\text{nm}}} = 4.25 \times 10^{-7}\ \text{m}$$

Now we can use Equation (7.1):

$$f = \frac{c}{\lambda} = \frac{3.0 \times 10^{8}\ \frac{\cancel{\text{m}}}{\text{sec}}}{4.25 \times 10^{-7}\ \cancel{\text{m}}} = \underline{7.1 \times 10^{14}\ \frac{1}{\text{sec}}}$$

4. Once again, we use Equation (7.1):

$$f = \frac{c}{\lambda}$$

But now we have to rearrange it so that we are solving for wavelength:

$$\lambda = \frac{c}{f}$$

Now we can plug in the numbers:

$$\lambda = \frac{3.0 \times 10^{8}\ \frac{\text{m}}{\text{s}}}{6.15 \times 10^{14}\ \frac{1}{\text{s}}} = \underline{4.9 \times 10^{-7}\ \text{m}}$$

5. $E = h \cdot f = (6.63 \times 10^{-34}\ \frac{J}{\cancel{Hz}}) \cdot (7.05 \times 10^{14}\ \cancel{Hz}) = \underline{4.67 \times 10^{-19}\ J}$

6. To get the energy, we start with the frequency:

$$f = \frac{c}{\lambda} = \frac{3.0 \times 10^{8}\ \frac{\cancel{m}}{sec}}{1.95 \times 10^{-7}\ \cancel{m}} = 1.5 \times 10^{15}\ \frac{1}{sec}$$

Now we can get the energy :

$$E = h \cdot f = (6.63 \times 10^{-34}\ \frac{J}{\cancel{Hz}}) \cdot (1.5 \times 10^{15}\ \cancel{Hz}) = \underline{9.9 \times 10^{-19}\ J}$$

7. We will have to start with the frequency:

$$E = h \cdot f$$

$$f = \frac{E}{h} = \frac{1.50 \times 10^{-18}\ \cancel{J}}{6.63 \times 10^{-34}\ \frac{\cancel{J}}{Hz}} = 2.26 \times 10^{15}\ Hz$$

Now we can get wavelength:

$$\lambda = \frac{3.0 \times 10^{8}\ \frac{m}{\cancel{s}}}{2.26 \times 10^{15}\ \frac{1}{\cancel{s}}} = \underline{1.3 \times 10^{-7}\ m}$$

8. a. To get to element Sc, we must go through row 1, which has two boxes in the s orbital block ($1s^2$). We then go through all of row 2 which has two boxes in the s orbital block and six boxes in the p orbital block ($2s^2 2p^6$). We also go through row 3, which has two boxes in the s orbital block and six in the p orbital block ($3s^2 3p^6$). We then go to the fourth row where we pass through both boxes in the s orbital block ($4s^2$). Finally, we go through one box in the d orbital block. Since we subtract one from the row number for d orbitals, this gives us $3d^1$. Thus, our final electron configuration is:

$$\underline{1s^2 2s^2 2p^6 3s^2 3p^6 4s^2 3d^1}$$

b. To get to element Cl, we must go through row 1, which has two boxes in the s orbital block ($1s^2$). We then go through all of row 2 which has two boxes in the s orbital block and six boxes in the p orbital block ($2s^2 2p^6$). We also go through both boxes in the s orbital block of row 3, ($3s^2$). Finally, we go through five boxes in the p orbital block of row 3, giving us $3p^5$. Thus, our final electron configuration is:

$$\underline{1s^2 2s^2 2p^6 3s^2 3p^5}$$

c. To get to element Ca, we must go through row 1, which has two boxes in the s orbital block ($1s^2$). We then go through all of row 2 which has two boxes in the s orbital block and six boxes in the p orbital block ($2s^2 2p^6$). We also go through row 3, which has two boxes in the s orbital block and six in the p orbital block ($3s^2 3p^6$). We then go to the fourth row where we go through the first two boxes, giving us $4s^2$. Thus, our final electron configuration is:

$$1s^2 2s^2 2p^6 3s^2 3p^6 4s^2$$

9. a. The nearest 8A element that has a lower atomic number than P is Ne. The only difference between P and Ne is that there are two boxes in the row 2, s orbital group and three boxes in the row 3, p orbital group. Thus, the abbreviated electron configuration for P is:

$$[Ne]3s^2 3p^3$$

b. The nearest 8A element that has a lower atomic number than Mo is Kr. The only difference between Mo and Kr is that there are two boxes in the row 5, s orbital group and four boxes in the row 4, d orbital group. Thus, the abbreviated electron configuration for Mo is:

$$[Kr]5s^2 4d^4$$

c. The nearest 8A element that has a lower atomic number than Ba is Xe. The only difference between Ba and Xe is that there are two boxes in the row 6, s orbital group. Thus, the abbreviated electron configuration for Ba is:

$$[Xe]6s^2$$

10 a. There are two many electrons in the 2s orbital. There should only be two electrons there.

b. The 3d orbital is not filled. You cannot go on to the next orbital until you fill up the one below.

c. The 4d orbital should be after the 5s orbital, not before it.

SOLUTIONS TO THE EXTRA PRACTICE PROBLEMS FOR MODULE #8

1. a. Sr is in group 2A, so it has 2 valence electrons: Sr

b. Ge is in group 4A, so it has 4 valence electrons: Ge

c. Te is in group 6A, so it has 6 valence electrons: S

2. a. Al is in group 3A, so it wants a charge of 3+. Oxygen is in group 6A, so oxide will have a charge of 2-. Ignoring the signs and switching the numbers gives us Al_2O_3.

b. Ca is in group 2A, so it wants a charge of 2+. Sulfur is in group 6A, so sulfide will have a charge of 2-. Since the charges are equal we ignore them, giving us CaS.

c. Cr is an exception, because there is a Roman numeral in the name. The numeral tells us that Cr want a charge of 3+. Chlorine is in group 7A, so chloride will have a charge of 1-. Ignoring the signs and switching the numbers gives us $CrCl_3$.

d. K is in group 1A, so it wants a charge of 1+. Nitrogen is in group 5A, so nitride will have a charge of 3-. Ignoring the signs and switching the numbers gives us K_3N.

3. Ionization potential decreases as you go down the chart. Thus, the order is Ra < Sr < Mg.

4. Electronegativity is a measure of an element's desire for electrons. It increases as you move to the right on the chart, so Br has the greatest desire for electrons.

5. Atomic radius decreases as you move to the right on the chart. The order is Ar < P < Al < Mg.

6. The chemical formula tells us that we have one N and three F's:

N F F F

This one is easy to put together, as N has three empty spaces and each F needs one more electron:

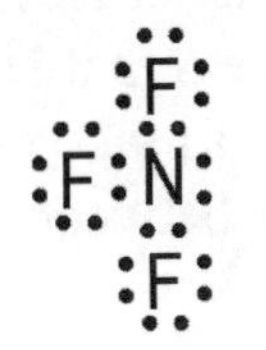

Now we just replace the shared electron pairs with bonds:

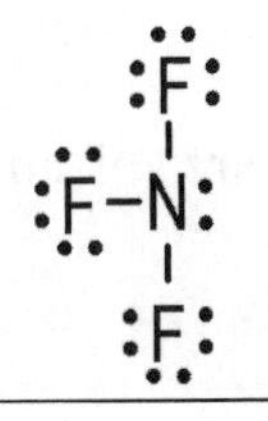

7. The chemical formula says that we have one C, two Cl's, and one S to play with:

·C· :Cl: :Cl: ·S:

The C has the most unpaired electrons and therefore goes in the middle:

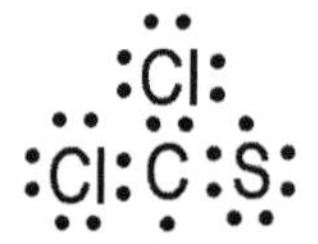

The Cl's have their ideal electron configurations, but the C and S are each one electron short. This can be fixed by taking the unpaired electrons and putting them between the C and S:

:Cl:

:Cl:C::S:

Now we just need to replace the shared electron pairs with bonds:

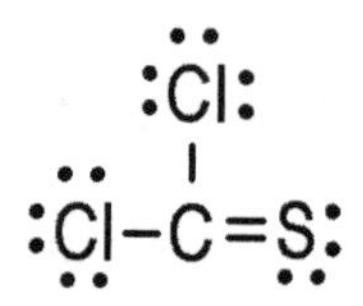

---

8. We have one Si and two S's to work with in this molecule:

·Si· ·S: ·S:

The Si will go in the middle, because it has the most unpaired electrons:

:S:Si:S:

No atom has its ideal electron configuration yet. However, if I take the unpaired electrons on the S atoms and put them each between the Si and the S that had the unpaired electron, then Si will have its ideal electron configuration. In the same way, if I take the unpaired electrons on the Si and put one between the Si and one of the S's and the other between Si and the other S, the S's will have their ideal electron configurations as well:

:S::Si::S:

Now we can just replace the shared electron pairs with bonds:

:S=Si=S:

9. In this molecule, we have one C and one S, so we might as well link them up:

·C:S:

Taking the unpaired electron on the S and putting it between the two atoms and doing the same with one of the unpaired electrons on the C gives S its ideal electron configuration:

·C::S:

The carbon still needs two more electrons. They must come from the S. Thus, we will put one of the electron pairs on the S in between the two atoms:

·C:::S:

Now both atoms have their ideal electron configuration. We should go ahead and pair up the two electrons on the C and replace the electron pairs with bonds:

:C≡S:

10. The formula tells us that we have one S and two Br's:

·S: :Br: :Br:

The S goes in the middle:

:Br:
:Br:S:

We now just put in the bonds:

:Br:
|
:Br–S:

11. Based on Lewis structures, the CS molecule would be the hardest to break, because it is held together with a triple bond.

SOLUTIONS TO THE EXTRA PRACTICE PROBLEMS FOR MODULE #9

1. Ionic compounds are named by simply listing the ions present. In order to get the formula, you must determine the charge of each ion and balance those charges.

a. The name indicates a sodium ion and a sulfate ion. Sodium is abbreviated with Na, and, since it is in group 1A, it has a charge of 1+. We are supposed to have memorized that the sulfate ion is SO_4 and has a charge of 2-. Ignoring the signs and switching the numbers gives us:

$\underline{Na_2SO_4}$

b. The name indicates a magnesium ion and a nitrate ion. Magnesium is abbreviated with Mg, and, since it is in group 2A, it has a charge of 2+. We are supposed to have memorized that the nitrate ion is NO_3 and has a charge of 1-. Ignoring the signs and switching the numbers gives us:

$\underline{Mg(NO_3)_2}$

c. The name indicates a calcium ion and a carbonate ion. Calcium is abbreviated with Ca, and, since it is in group 2A, it has a charge of 2+. We are supposed to have memorized that the carbonate ion is CO_3 and has a charge of 2-. Since the numerical values of the charges are the same, we ignore them.

$\underline{CaCO_3}$

d. The name indicates an aluminum ion and a chromate ion. Aluminum is abbreviated with an Al, and, since it is in group 3A, it has a charge of 3+. We are supposed to have memorized that the chromate ion is CrO_4 and has a charge of 2-. Ignoring the signs and switching the numbers gives us:

$\underline{Al_2(CrO_4)_3}$

2. In order to name ionic compounds, we only have to put the names of the ions together.

a. Since we see that NH_4 is in parentheses, that means it is a polyatomic ion. We are supposed to have memorized that NH_4^+ is the ammonium ion, and the only other ion is the single-atom sulfide ion. Thus, the name is <u>ammonium sulfide</u>.

b. In looking at this molecule, we should notice the NO_3. It tells us the nitrate polyatomic ion is present. The only thing left after that is the potassium ion. Thus, the name is <u>potassium nitrate</u>.

c. Since we see that PO_4 is in parentheses, that means it is a polyatomic ion. We are supposed to have memorized that PO_4^{3-} is the phosphate ion, and the only other ion is the single-atom magnesium ion. Thus, the name is <u>magnesium phosphate</u>.

d. In looking at this molecule, we should notice the PO_4. It tells us that the phosphate polyatomic ion is present. The only thing left after that is the aluminum ion. Thus, the name is <u>aluminum phosphate</u>.

3. This molecule has only two atoms. Thus, it is <u>linear with a bond angle of 180 degrees</u>. The picture is given to the right.

:C≡S:

4. We start with the Lewis structure:

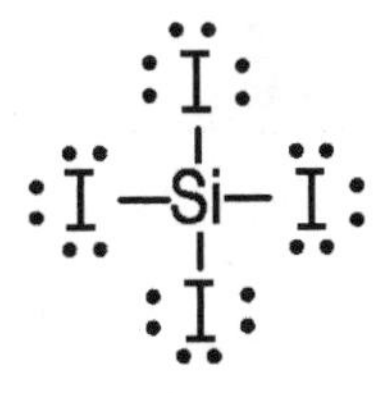

The central atom has four groups of electrons around it, and non of them are nonbonding electron pairs. Thus, the shape is tetrahedral with a bond angle of 109 degrees:

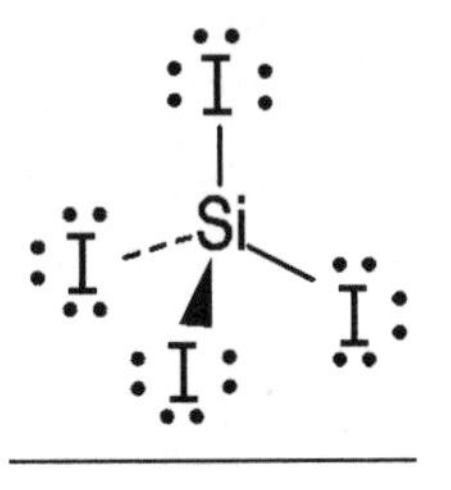

5. We start with the Lewis Structure:

The central atom has four groups of electrons around it, so the base geometry is a tetrahedron. However, one "leg" of the tetrahedron is missing, because one of the groups is not a bond. Thus, this molecule is pyramidal with a bond angle of 107 degrees:

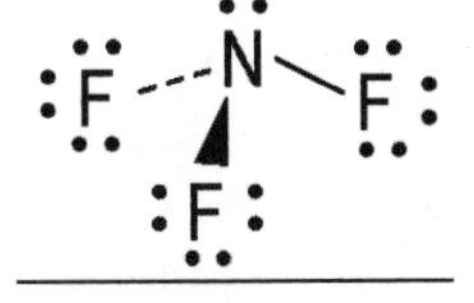

6. We start with the Lewis structure:

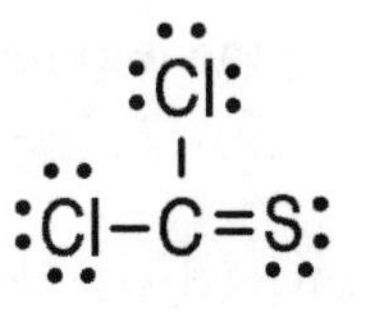

The central atom has three groups (three bonds) around it. One of those bonds is a double bond, but it still counts as only one bond. Thus, this molecule is trigonal with a bond angle of 120 degrees.

7. We start with the Lewis structure:

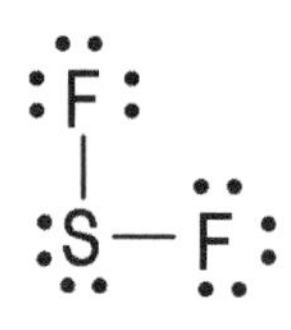

The central atom has four groups of electrons around it, so the base geometry is tetrahedral. However, two of the "legs" in the tetrahedron are gone, so the resulting structure is bent with a bond angle of 105 degrees:

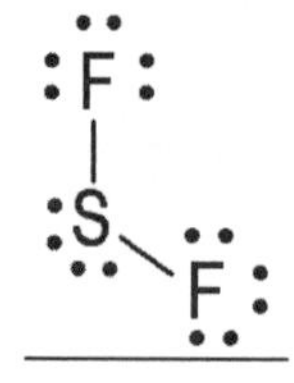

8. To be polar covalent, the molecule must have polar covalent bonds and those bonds cannot cancel each other out. All of the molecules under consideration have polar covalent bonds. However, in the tetrahedral geometry of SiI_4, those bonds cancel. The bonds do not cancel in the other molecules, however, so all of the molecules except SiI_4 are polar covalent, while SiI_4 is purely covalent.

9. a. This molecule has no metals, so it is not ionic. If you determine its geometry, you will find it is tetrahedral. Usually, polar covalent bonds cancel in a tetrahedral geometry. This is only true, however, if the bonds are identical. In this molecule, they are not. Thus, the bonds do not cancel, and the result is a polar covalent molecule.

b. This molecule has no metals, so it is not ionic. It has polar bonds, but the Lewis structure has only two groups of electrons around the central atom. Thus, this molecule is linear, with an S on either side of the C. Since the bonds pull opposite one another, they cancel, making this a purely covalent molecule.

c. This molecule has a metal (Al), so it is ionic.

d. This molecule is purely covalent, since it has no polar covalent bonds.

e. This molecule has no metals, so it is not ionic. If you determine its geometry, you will find it is tetrahedral. Since the bonds are all identical in the tetrahedron, they cancel, making this a purely covalent molecule.

f. This molecule has no metals, so it is not ionic. If you determine its geometry, you will find it is pyramidal. There is no way for the bonds to cancel in a pyramidal structure, so the result is a polar covalent molecule.

10. The ionic and polar covalent molecules should dissolve in water. Thus, you would expect $SiFCl_3$, AlF_3, and PF_3 to dissolve in water.

SOLUTIONS TO THE EXTRA PRACTICE PROBLEMS FOR MODULE #10

1. H_2CO_3 is the acid, because it lost a H^+ to become HCO_3^-. That makes water the base, because it accepted an H^+ to become H_3O^+.

2. KOH is the base, because when its ions split apart, the hydroxide ion (OH^-) accepted an H^+ to become water. This make C_2H_6O the acid, since it lost an H^+ to become $C_2H_5O^-$, which became a part of the salt.

3. $Mg(OH)_2$ is the base, because when its ions split apart, the hydroxide ion (OH^-) accepted an H^+ to become water. This make HCl the acid, since it lost an H^+ to become Cl^-, which became a part of the salt.

4. This is an acid and an ionic base, so it will follow the acid + base → salt + water formula. The salt comes from the metal ion of the base (Al^{3+}) and the negative ion left over when the acid donates it H^+ ions (I^-). To determine the formula, you switch the charges and drop the signs to make AlI_3. Thus, the unbalanced equation is:

$$HI + Al(OH)_3 \rightarrow AlI_3 + H_2O$$

Balancing gives us:

$$\underline{3HI + Al(OH)_3 \rightarrow AlI_3 + 3H_2O}$$

5. This is an acid and an ionic base, so it will follow the acid + base → salt + water formula. The salt comes from the metal ion of the base (Na^+) and the negative ion left over when the acid donates it H^+ ions. This acid need to donate three H^+ ions, and when that is done, PO_4^{3-} is left behind. To determine the formula, you switch the charges and drop the signs to make Na_3PO_4. Thus, the unbalanced equation is:

$$H_3PO_4 + NaOH \rightarrow Na_3PO_4 + H_2O$$

Balancing gives us:

$$\underline{H_3PO_4 + 3NaOH \rightarrow Na_3PO_4 + 3H_2O}$$

6. Since this involves a covalent base, we cannot use the acid + base → salt + water formula. Instead, we just have to rely on the definitions of acids and bases. Acids donate H^+, and bases accept H^+. H_3PO_4 wants to donate three H^+ ions, but the base can accept only one, so it takes three bases to get the job done:

$$\underline{H_3PO_4 + 3CH_5N \rightarrow PO_4^{3-} + 3CH_6N^+}$$

7. a. Molarity is given by number of moles divided by number of liters. We have both those units, so we just divide them:

$$\text{Concentration} = \frac{\#\text{ moles}}{\#\text{ liters}} = \frac{1.11\text{ moles HCl}}{1.5\text{ L}} = \underline{0.74\text{ M}}$$

b. In order to get concentration, we must have moles and liters. The problem gives us grams and mL, so we must make two conversions:

$$\frac{23.1 \cancel{\text{g NaOH}}}{1} \times \frac{1 \text{ mole NaOH}}{40.0 \cancel{\text{g NaOH}}} = 0.578 \text{ moles NaOH}$$

$$\frac{500.0 \cancel{\text{mL}}}{1} \times \frac{0.001 \text{ L}}{1 \cancel{\text{mL}}} = 0.5000 \text{ L}$$

Now we can calculate molarity:

$$\text{Concentration} = \frac{\#\text{moles}}{\#\text{liters}} = \frac{0.578 \text{ moles NaOH}}{0.5000 \text{ L}} = 1.16 \frac{\text{moles NaOH}}{\text{L}} = \underline{1.16 \text{ M}}$$

c. In order to get concentration, we must have moles and liters. The problem gives us grams and mL, so we must make two conversions:

$$\frac{14.5 \cancel{\text{g H}_2\text{CO}_3}}{1} \times \frac{1 \text{ mole H}_2\text{CO}_3}{62.0 \cancel{\text{g H}_2\text{CO}_3}} = 0.234 \text{ moles H}_2\text{CO}_3$$

$$\frac{200.0 \cancel{\text{mL}}}{1} \times \frac{0.001 \text{ L}}{1 \cancel{\text{mL}}} = 0.2000 \text{ L}$$

Now we can calculate molarity:

$$\text{Concentration} = \frac{\#\text{moles}}{\#\text{liters}} = \frac{0.234 \text{ moles H}_2\text{CO}_3}{0.2000 \text{ L}} = 1.17 \frac{\text{moles H}_2\text{CO}_3}{\text{L}} = \underline{1.17 \text{ M}}$$

8. This is a dilution problem, so we use the dilution equation. M_1 is 6.78 M, we need to determine V_1. M_2 is 1.15 M and $V_2 = 500.0$ mL.

$$M_1V_1 = M_2V_2$$

$$(6.78 \text{ M}) \cdot V_1 = (1.15 \text{ M}) \cdot (500.0 \text{ mL})$$

$$V_1 = \frac{1.15 \cancel{\text{M}} \cdot 500.0 \text{ mL}}{6.78 \cancel{\text{M}}} = 84.8 \text{ mL}$$

The chemist must take 84.8 mL of the original solution and dilute it with enough water to make 500.0 mL of solution.

9. This is a dilution problem, so we use the dilution equation. M_1 is 5.11 M, we need to determine V_1. M_2 is 4.5 M and $V_2 = 50.0$ mL.

$$M_1V_1 = M_2V_2$$

$$(5.11 \text{ M}) \cdot V_1 = (4.5 \text{ M}) \cdot (50.0 \text{ mL})$$

$$V_1 = \frac{4.5 \cancel{\text{M}} \cdot 50.0 \text{ mL}}{5.11 \cancel{\text{M}}} = 44 \text{ mL}$$

<u>You must take 44 mL of the original solution and dilute it with enough water to make 50.0 mL of solution</u>.

10. Remember, titrations are just stoichiometry problems, so first we have to come up with a balanced chemical equation:

$$HBr + NaOH \rightarrow NaBr + H_2O$$

Since the endpoint was reached, we know that there was exactly enough base added to eat up all of the acid. First, then, we calculate how many moles of base were added:

$$\frac{1.14 \text{ moles NaOH}}{1 \cancel{\text{L}}} \times \frac{0.0450 \cancel{\text{L}}}{1} = 0.0513 \text{ moles NaOH}$$

We can now use the chemical equation to determine how many moles of acid were present:

$$\frac{0.0513 \cancel{\text{moles NaOH}}}{1} \times \frac{1 \text{ mole HBr}}{1 \cancel{\text{mole NaOH}}} = 0.0513 \text{ moles HBr}$$

Now that we have the number of moles of acid present, we simply divide by the volume of acid to get concentration:

$$\text{Concentration} = \frac{\#\text{ moles}}{\#\text{ liters}} = \frac{0.0513 \text{ moles HBr}}{0.0150 \text{ L}} = \underline{3.42 \text{ M}}$$

11. First we have to come up with a balanced chemical equation:

$$2HCl + Mg(OH)_2 \rightarrow MgCl_2 + 2H_2O$$

Since the endpoint was reached, we know that there was exactly enough acid added to eat up all of the base. First, then, we calculate how many moles of acid were added:

$$\frac{1.00 \text{ mole HCl}}{1 \cancel{\text{L}}} \times \frac{0.00351 \cancel{\text{L}}}{1} = 0.00351 \text{ moles HCl}$$

We can now use the chemical equation to determine how many moles of base were present:

$$\frac{0.00351 \cancel{\text{moles HCl}}}{1} \times \frac{1 \text{ mole Mg(OH)}_2}{2 \cancel{\text{moles HCl}}} = 0.00176 \text{ moles Mg(OH)}_2$$

Now that we have the number of moles of base present, we simply divide by the volume of acid to get concentration:

$$\text{Concentration} = \frac{\text{\# moles}}{\text{\# liters}} = \frac{0.00176 \text{ moles Mg(OH)}_2}{0.0500 \text{ L}} = \underline{0.0352 \text{ M}}$$

SOLUTIONS TO THE EXTRA PRACTICE PROBLEMS FOR MODULE #11

1. The solubility of gases increases with decreasing temperature. Thus, you should cool the liquid.

2. This is just a stoichiometry problem. We can tell this by the fact that we are being asked to determine the amount of one substance when we are given the amount of another substance. The only way to do that is by stoichiometry. Now, in order to do stoichiometry, we must first get our amount in moles.

$$\frac{1.25 \text{ moles } Al(NO_3)_3}{1 \cancel{L}} \times 0.191 \cancel{L} = 0.239 \text{ moles } Al(NO_3)_3$$

Now that we have moles, we can do stoichiometry:

$$\frac{0.239 \cancel{\text{moles } Al(NO_3)_3}}{1} \times \frac{1 \text{ mole } Al_2(CO_3)_3}{2 \cancel{\text{moles } Al(NO_3)_3}} = 0.120 \text{ moles } Al_2(CO_3)_3$$

Now, of course, this is not quite the answer we need. We were asked to figure out how many grams of aluminum carbonate were produced, so we have to convert from moles back to grams:

$$\frac{0.120 \cancel{\text{moles } Al_2(CO_3)_3}}{1} \times \frac{234.0 \text{ grams } Al_2(CO_3)_3}{1 \cancel{\text{mole } Al_2(CO_3)_3}} = \underline{28.1 \text{ grams } Al_2(CO_3)_3}$$

3. We must first get our amount in moles.

$$\frac{50.0 \cancel{\text{g KCN}}}{1} \times \frac{1 \text{ mole KCN}}{65.1 \cancel{\text{g KCN}}} = 0.768 \text{ moles KCN}$$

Now that we have moles, we can do stoichiometry:

$$\frac{0.768 \cancel{\text{moles KCN}}}{1} \times \frac{1 \text{ mole HCN}}{1 \cancel{\text{mole KCN}}} = 0.768 \text{ moles HCN}$$

Now, of course, this is not quite the answer we need. We were asked to figure out how many mL of the HCN solution is needed :

$$\frac{0.768 \cancel{\text{moles HCN}}}{1} \times \frac{1 \text{ L}}{1.51 \cancel{\text{moles HCN}}} = 0.509 \text{ L} = \underline{509 \text{ mL}}$$

4. We must first get our amount in moles.

$$\frac{50.0 \cancel{\text{g Cu}}}{1} \times \frac{1 \text{ mole Cu}}{63.5 \cancel{\text{g Cu}}} = 0.787 \text{ moles Cu}$$

Now that we have moles, we can do stoichiometry:

$$\frac{0.787\ \cancel{\text{moles Cu}}}{1} \times \frac{8\ \text{moles}\ HNO_3}{3\ \cancel{\text{moles Cu}}} = 2.10\ \text{moles}\ HNO_3$$

Now, of course, this is not quite the answer we need. We were asked to figure out how many mL of the HNO_3 solution is needed :

$$\frac{2.10\ \cancel{\text{moles}\ HNO_3}}{1} \times \frac{1\ \text{L}}{3.5\ \cancel{\text{moles}\ HNO_3}} = 0.600\ \text{L} = \underline{6.00 \times 10^2\ \text{mL}}$$

5. To calculate molality, we must have moles of solute and kg of solvent .

$$\frac{50.0\ \cancel{\text{g}\ Mg(NO_3)_2}}{1} \times \frac{1\ \text{mole}\ Mg(NO_3)_2}{148.3\ \cancel{\text{g}\ Mg(NO_3)_2}} = 0.337\ \text{moles}\ Mg(NO_3)_2$$

$$\frac{500.0\ \cancel{\text{g}}}{1} \times \frac{1\ \text{kg}}{1{,}000\ \cancel{\text{g}}} = 0.5000\ \text{kg}$$

Now that we have moles of solute and kg of solvent, we can use Equation (11.1):

$$\text{molality} = \frac{\#\ \text{moles solute}}{\#\ \text{kg solvent}} = \frac{0.337\ \text{moles}\ Mg(NO_3)_2}{0.5000\ \text{kg water}} = \underline{0.674\ \text{m}}$$

6. First, we need to see how many moles of $CaCl_2$ to add:

$$\text{molality} = \frac{\#\ \text{moles solute}}{\#\ \text{kg solvent}}$$

$$\frac{\text{moles}\ CaCl_2}{0.125\ \text{kg water}} = \underline{2.0\ \text{m}}$$

$$\text{moles}\ CaCl_2 = 0.25$$

Thus, we have to add 0.25 moles of $CaCl_2$ to 125 kg of water to make a 2.0 m solution. Now we just need to see how many grams that is:

$$\frac{0.25\ \cancel{\text{moles}\ CaCl_2}}{1} \times \frac{111.1\ \text{g}\ CaCl_2}{1\ \cancel{\text{mole}\ CaCl_2}} = \underline{28\ \text{g}\ CaCl_2}$$

7. You want the solute that splits up into the most ions. $\underline{Ca_3(PO_4)_2}$ splits into 5 ions (three calcium ions and two phosphate ions. That is more than the other two, so it would give the lowest freezing point to water.

8. In a freezing-point depression problem, you must use Equation (11.2). However, in order to use that equation, we must know K_f, i, and m. Right now, we only know K_f. However, we have been given enough information to calculate both "i" and "m." First let's calculate m:

$$\frac{10.0\ \cancel{\text{g KF}}}{1} \times \frac{1\ \text{mole KF}}{58.1\ \cancel{\text{g KF}}} = 0.172\ \text{moles KF}$$

$$\text{m} = \frac{0.172\ \text{moles KF}}{0.1000\ \text{kg water}} = 1.72\ \text{m}$$

To figure out "i", we just have to realize that according to its formula, KF splits up into one potassium ion and one fluoride ion. Thus, i = 2. Now that we have all of the components of Equation (11.2), we can use it:

$$\Delta T = -\text{i} \cdot K_f \cdot \text{m} = -2 \cdot 1.86 \frac{^\circ\text{C}}{\cancel{\text{m}}} \cdot 1.72\ \cancel{\text{m}} = -6.40\ ^\circ\text{C}$$

So the freezing point is 6.40 °C lower than that of normal water, or <u>-6.40 °C</u>.

9. We can use Equation (11.2) to solve this. Since $CaCl_2$ is made up of one calcium ion and two chloride ions, i = 3.

$$\Delta T = -\text{i} \cdot K_f \cdot \text{m}$$

$$\text{m} = \frac{\Delta T}{-\text{i} \cdot K_f} = \frac{-5.00\ ^\circ\cancel{\text{C}}}{-3 \cdot 1.86 \frac{^\circ\cancel{\text{C}}}{\text{molal}}} = \underline{0.896\ \text{molal}}$$

10. To calculate boiling points, we must use Equation (11.3). To do that, however, we must know "i" and "m". To calculate "m":

$$\frac{100.0\ \cancel{\text{g (NH}_4\text{)}_2\text{S}}}{1} \times \frac{1\ \text{mole (NH}_4\text{)}_2\text{S}}{68.2\ \cancel{\text{g (NH}_4\text{)}_2\text{S}}} = 1.47\ \text{moles} (\text{NH}_4)_2\text{S}$$

$$\text{m} = \frac{1.47\ \text{moles} (\text{NH}_4)_2\text{S}}{0.7500\ \text{kg water}} = 1.96\ \text{m}$$

Since ammonium sulfide is an ionic compound, it dissolves by splitting up into its two ammonium ions and its one sulfide ion. Thus, i = 3.

$$\Delta T = \text{i} \cdot K_b \cdot \text{m} = 3 \cdot 0.512 \frac{^\circ\text{C}}{\cancel{\text{m}}} \cdot 1.96\ \cancel{\text{m}} = 3.01\ ^\circ\text{C}$$

This means that the boiling point of the solution is 3.01 °C *higher* than that of pure water. The boiling point of pure water is 100.0 °C, so the boiling point of this solution is <u>103.0 °C</u>.

SOLUTIONS TO THE EXTRA PRACTICE PROBLEMS FOR MODULE #12

1. This problem asks you to predict how a gas will change when you change some of the conditions under which it is stored. This means that you need to use the combined gas law (Equation 12.10).

$$\frac{P_1V_1}{T_1} = \frac{P_2V_2}{T_2}$$

According to this problem, $T_1 = 298.2$ K, $V_1 = 5.6$ L, and $T_2 = 77.2$ K. Also, the problem states that the pressure does not change; thus, P_1 and P_2 cancel out:

$$\frac{\cancel{P_1}V_1}{T_1} = \frac{\cancel{P_2}V_2}{T_2}$$

We can now rearrange the equation to solve for the new volume:

$$\frac{V_1 \cdot T_2}{T_1} = V_2$$

Now we can put in the numbers and determine the new volume:

$$\frac{5.6\text{ L} \cdot 77.2\ \cancel{\text{K}}}{298.2\ \cancel{\text{K}}} = \underline{1.4\text{ L}}$$

2. This is obviously another combined gas law problem, with $P_1 = 755$ mmHg, $V_1 = 6.65$ L, $T_1 = 298.2$ K, $P_2 = 625$ mmHg, and $T_2 = 288.2$ K. The problem asks us to determine the new volume, so we have to rearrange Equation (12.10) to solve for V_2:

$$\frac{P_1V_1T_2}{T_1P_2} = V_2$$

Now we can plug in the numbers:

$$V_2 = \frac{755\ \cancel{\text{mmHg}} \cdot 6.65\text{ L} \cdot 288.2\ \cancel{\text{K}}}{298.2\ \cancel{\text{K}} \cdot 625\ \cancel{\text{mmHg}}} = \underline{7.76\text{ L}}$$

3. This is obviously another combined gas law problem, with $P_1 = 740$ torr, $V_1 = 56.7$ mL, $T_1 = 298$ K, $P_2 = 1.00$ atm (standard pressure), and $T_2 = 273$ K (standard temperature). The problem asks us to determine the new volume, so we have to rearrange Equation (12.10) to solve for V_2:

$$\frac{P_1V_1T_2}{T_1P_2} = V_2$$

We need to make the pressure units the same. We can do this by converting torr into atm or vice-versa. I will choose to do the latter:

$$P_2 = \frac{1.00\ \cancel{\text{atm}}}{1} \times \frac{760\text{ torr}}{1\ \cancel{\text{atm}}} = 7.60 \times 10^2\text{ torr}$$

Now we can plug in the numbers:

$$V_2 = \frac{740\ \cancel{\text{torr}} \cdot 56.7\ \text{mL} \cdot 273\ \cancel{\text{K}}}{298\ \cancel{\text{K}} \cdot 7.60 \times 10^2\ \cancel{\text{torr}}} = \underline{51\ \text{mL}}$$

4. The total pressure is just the sum of the partial pressures. Thus, if the total pressure is 1.0 atm, the partial pressure of oxygen is:

$$1.0\ \text{atm} - 0.7\ \text{atm} = \underline{0.3\ \text{atm}}$$

5. To determine partial pressure from total pressure, we need to know the mole fractions involved:

$$\frac{0.200\ \cancel{\text{g}H_2}}{1} \times \frac{1\ \text{mole}\ H_2}{2.02\ \cancel{\text{g}H_2}} = 0.0990\ \text{moles}\ H_2$$

$$\frac{4.00\ \cancel{\text{g}CO_2}}{1} \times \frac{1\ \text{mole}\ CO_2}{44.0\ \cancel{\text{g}CO_2}} = 0.0909\ \text{moles}\ CO_2$$

$$\frac{1.45\ \cancel{\text{g}N_2}}{1} \times \frac{1\ \text{mole}\ N_2}{28.0\ \cancel{\text{g}N_2}} = 0.0518\ \text{moles}\ N_2$$

Now that we have the number of moles of each component, we can calculate the total number of moles in the mixture:

$$\text{Total number of moles} = 0.0990\ \text{moles} + 0.0909\ \text{moles} + 0.0518\ \text{moles} = 0.2417\ \text{moles}$$

Plugging that into Equation (12.12):

$$X_{H_2} = \frac{0.0990\ \cancel{\text{moles}}}{0.2417\ \cancel{\text{moles}}} = 0.410$$

$$X_{CO_2} = \frac{0.0909\ \cancel{\text{moles}}}{0.2417\ \cancel{\text{moles}}} = 0.376$$

$$X_{N_2} = \frac{0.0518\ \cancel{\text{moles}}}{0.2417\ \cancel{\text{moles}}} = 0.214$$

Using the mole fractions we just obtained:

$$P_{H_2} = 0.410 \cdot 1.5\ \text{atm} = \underline{0.62\text{atm}}$$

$$P_{CO_2} = 0.376 \cdot 1.5\ \text{atm} = \underline{0.56\text{atm}}$$

$$P_{N_2} = 0.214 \cdot 1.5\ \text{atm} = \underline{0.32\ \text{atm}}$$

6. In this problem, we are given pressure and temperature and the mass (from which we can get moles). We are then asked to calculate V. We can do this by rearranging the ideal gas law:

$$PV = nRT$$

$$V = \frac{nRT}{P}$$

Now we just need to get "n":

$$\frac{12.1\ \cancel{g\ N_2}}{1} \times \frac{1\text{ mole } N_2}{28.0\ \cancel{g\ N_2}} = 0.432\text{ moles } N_2$$

Now we can use the equation, remembering to convert temperature to K:

$$V = \frac{0.432\ \cancel{\text{moles}} \cdot 0.0821 \frac{L \cdot \cancel{\text{atm}}}{\cancel{\text{mole} \cdot K}} \cdot 273\ \cancel{K}}{1.00\ \cancel{\text{atm}}} = \underline{9.68\ L}$$

7. In this problem, we are given temperature and volume and the mass (from which we can get moles). We are then asked to calculate P. We can do this by rearranging the ideal gas law:

$$PV = nRT$$

$$P = \frac{nRT}{V}$$

Now we just need to get "n":

$$\frac{5.0\ \cancel{g\ N_2}}{1} \times \frac{1\text{ mole } N_2}{28.0\ \cancel{g\ N_2}} = 0.18\text{ moles } N_2$$

Now we can use the equation, remembering to convert temperature to K:

$$P = \frac{0.18\ \cancel{\text{moles}} \cdot 0.0821 \frac{\cancel{L} \cdot \text{atm}}{\cancel{\text{mole} \cdot K}} \cdot 294.2\ \cancel{K}}{1.00\ \cancel{L}} = \underline{4.3\text{ atm}}$$

8. In this stoichiometry problem, we are given the amount of limiting reactant and asked to calculate how much product will be made. We start by converting the amount of limiting reactant to moles:

$$\frac{110.0\ \cancel{g\ CS_2}}{1} \times \frac{1\text{ mole } CS_2}{76.2\ \cancel{g\ CS_2}} = 1.44\text{ moles } CS_2$$

We can then use stoichiometry to determine the number of moles of SO_2 produced:

$$\frac{1.44 \text{ } \cancel{\text{moles}CS_2}}{1} \times \frac{2 \text{ moles} SO_2}{1 \text{ } \cancel{\text{mole}CS_2}} = 2.88 \text{ moles} SO_2$$

Now we need to use the ideal gas law, realizing that we must convert the temperature to Kelvin:

$$PV = nRT$$

$$V = \frac{nRT}{P} = \frac{2.88 \text{ } \cancel{\text{moles}} \cdot 0.0821 \frac{L \cdot \cancel{atm}}{\cancel{mole} \cdot \cancel{K}} \cdot 614 \cancel{K}}{2.1 \text{ } \cancel{atm}} = \underline{69 \text{ L}}$$

9. We start by converting the amount of limiting reactant to moles:

$$\frac{100.0 \text{ } \cancel{gTiO_2}}{1} \times \frac{1 \text{ mole} TiO_2}{79.9 \text{ } \cancel{gTiO_2}} = 1.25 \text{ moles} TiO_2$$

We can then use stoichiometry to determine the number of moles of Cl_2 used:

$$\frac{1.25 \text{ } \cancel{\text{moles}TiO_2}}{1} \times \frac{6 \text{ moles} Cl_2}{3 \text{ } \cancel{\text{moles}TiO_2}} = 2.50 \text{ moles} Cl_2$$

Now we need to use the ideal gas law, realizing that we must convert the temperature to Kelvin:

$$PV = nRT$$

$$V = \frac{nRT}{P} = \frac{2.50 \text{ } \cancel{\text{moles}} \cdot 0.0821 \frac{L \cdot \cancel{atm}}{\cancel{mole} \cdot \cancel{K}} \cdot 273 \cancel{K}}{1.00 \text{ } \cancel{atm}} = \underline{56.0 \text{ L}}$$

10. We start by converting the amount of limiting reactant to moles:

$$\frac{100.0 \text{ } \cancel{gH_2SO_4}}{1} \times \frac{1 \text{ mole} H_2SO_4}{98.1 \text{ } \cancel{gH_2SO_4}} = 1.02 \text{ moles} H_2SO_4$$

We can then use stoichiometry to determine the number of moles of HCl produced:

$$\frac{1.02 \text{ } \cancel{\text{moles}H_2SO_4}}{1} \times \frac{1 \text{ mole HCl}}{1 \text{ } \cancel{\text{mole}H_2SO_4}} = 1.02 \text{ moles HCl}$$

Finishing:

$$PV = nRT$$

$$V = \frac{nRT}{P} = \frac{1.02 \text{ } \cancel{\text{moles}} \cdot 0.0821 \frac{L \cdot \cancel{atm}}{\cancel{mole} \cdot \cancel{K}} \cdot 623.2 \cancel{K}}{1.00 \text{ } \cancel{atm}} = \underline{52.2 \text{ L}}$$

SOLUTIONS TO THE EXTRA PRACTICE PROBLEMS FOR MODULE #13

1. Elements in their elemental form have a ΔH_f^o of zero. Only F_2 (g) and O_2 (g) fit the bill.

2. Endothermic reactions have products at a higher energy than reactants. Thus, diagram II represents an endothermic reaction. The ΔH is roughly 55 kcals, which is the difference between the energy of the products and that of the reactants. Your number can range from 50 to 60, since it is being read from a graph.

3. The activation energy is given by the difference in energy between the intermediate state and the reactants. Thus, diagram I represents the reaction with the lowest activation energy.

4. Yes, it is possible for the reaction to occur. In order for the reaction to occur, however, heat must be released, and that must disorder the surroundings more than the chemicals become ordered.

5. a. There is one gas molecule on the reactants side and none on the products side. ΔS is negative.

b. There are no gas molecules on the reactants side and one on the products side. ΔS is positive.

c. There are nine gas molecules on the reactants side and 10 on the products side. ΔS is positive.

d. There are no gas molecules at all, so we look at liquid and aqueous states. There are three aqueous molecules on the reactants side and one on the products side. ΔS is negative.

6. To use bond energies, we have to draw the Lewis structures:

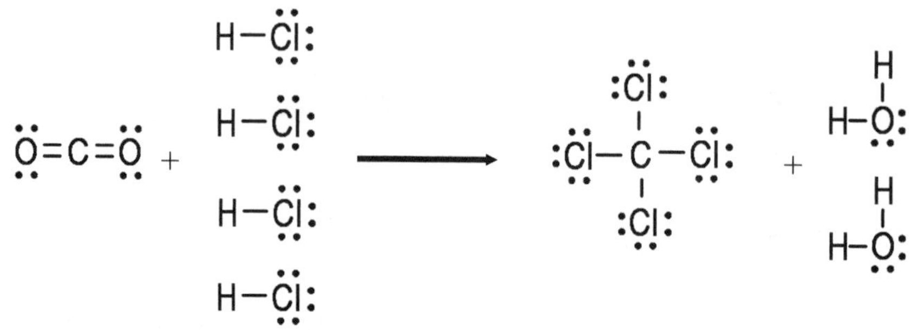

Now we can see what bonds need to be broken and what ones need to be formed. We need to break two C=O bonds (each worth of 799 kJ/mole) and four H-Cl bonds (each worth 428 kJ/mole). In the process, four C-Cl bonds (worth 327 kJ/mole each) and four H-O bonds (worth 459 kJ/moles each) are formed. Equation (13.6), then, looks like this:

$$\Delta H = (2 \text{ moles}) \times (799 \frac{\text{kJ}}{\text{mole}}) + (4 \text{ moles}) \times (428 \frac{\text{kJ}}{\text{mole}}) - (4 \text{ moles}) \times (327 \frac{\text{kJ}}{\text{mole}}) - (4 \text{ moles}) \times (459 \frac{\text{kJ}}{\text{mole}})$$

$$\Delta H = \underline{166 \text{ kJ}}$$

7. The ΔH_f^o of O_2 (g) is zero. The rest are in Table 13.2:

$$\Delta H^\circ = (4 \cancel{\text{moles}}) \times (90.3 \frac{\text{kJ}}{\cancel{\text{mole}}}) + (6 \cancel{\text{moles}}) \times (-242 \frac{\text{kJ}}{\cancel{\text{mole}}}) - (4 \cancel{\text{moles}}) \times (-45.9 \frac{\text{kJ}}{\cancel{\text{mole}}})$$

$$\Delta H^\circ = \underline{-907 \text{ kJ}}$$

8. We need to use Table 13.3 to solve this:

$$\Delta S^\circ = (4 \cancel{\text{moles}}) \times (211 \frac{\text{J}}{\cancel{\text{mole}} \cdot \text{K}}) + (6 \cancel{\text{moles}}) \times (189 \frac{\text{J}}{\cancel{\text{mole}} \cdot \text{K}}) - (4 \cancel{\text{moles}}) \times (193 \frac{\text{J}}{\cancel{\text{mole}} \cdot \text{K}})$$
$$- (5 \cancel{\text{moles}}) \times (205.0 \frac{\text{J}}{\cancel{\text{mole}} \cdot \text{K}})$$

$$\Delta S^\circ = \underline{181 \frac{\text{J}}{\text{K}}}$$

9. We need to see for what temperatures ΔG is negative. Thus:

$$\Delta H - T\Delta S < 0$$

In order to use this equation, though, we need to get our units consistent:

$$\frac{-114 \cancel{\text{J}}}{\text{mole} \cdot \text{K}} \times \frac{1 \text{ kJ}}{1{,}000 \cancel{\text{J}}} = -0.114 \frac{\text{kJ}}{\text{mole} \cdot \text{K}}$$

Now we can use the equation:

$$-623 \frac{\text{kJ}}{\text{mole}} - \text{T} \cdot (-0.114 \frac{\text{kJ}}{\text{mole} \cdot \text{K}}) < 0$$

We can solve this equation like any algebraic equation. We just need to remember that if we divide or multiply by a negative number, we must reverse the inequality sign:

$$\text{T} \cdot (0.114 \frac{\text{kJ}}{\text{mole} \cdot \text{K}}) < 623 \frac{\text{kJ}}{\text{mole}}$$

$$\text{T} < \frac{623 \frac{\cancel{\text{kJ}}}{\cancel{\text{mole}}}}{0.114 \frac{\cancel{\text{kJ}}}{\cancel{\text{mole}} \cdot \text{K}}}$$

$$\underline{\text{T} < 5.46 \times 10^3 \text{ K}}$$

SOLUTIONS TO THE EXTRA PRACTICE PROBLEMS FOR MODULE #14

1. Reaction rate has units of M/s. When solving for the rate constant, you divide by the concentrations of the reactants raised to their orders. To get these units for the rate constant, then, the rate must be divided by M^2. That means the overall order is 2.

2. If the order is three, then you must raise the concentration to the third power. If you double the concentration, that would be like raising 2 to the third power, which is 8. Thus, the rate increases by a factor of 8.

3. The rate equation for this reaction will look like:

$$R = k[NO]^x[Cl_2]^y$$

To figure out k, x, and y, we have to look at the data from the experiment. The value for x can be determined by comparing two trials in which the concentration of NO changes, but the concentration of Cl_2 stays the same. This would correspond to trials 1 and 2. In these two trials, the concentration of NO doubled, and the rate went up by a factor of 4. This means that x = 2, because the only way you can get a 4-fold increase in rate from a doubling of the concentration is by squaring the concentration. The value for y can be determined by looking at trials 2 and 3, where the concentration of NO stayed the same but the concentration of O_2 doubled. When that happened, the rate increased by a factor of 2. Since rate doubled when concentration doubled, that means y = 1. Thus, the rate equation becomes:

$$R = k[NO]^2[Cl_2]$$

Now that we have x and y, we only need to find out the value for k. We can do this by using any one of the trials in the experiment and plugging the data into the equation. The only unknown will be k, and we can therefore solve for it:

$$R = k[NO]^2[Cl_2]$$

$$0.113\frac{M}{s} = k \cdot (0.050\ M)^2 \cdot (0.050\ M)$$

$$k = \frac{0.113\ \frac{\cancel{M}}{s}}{(0.0025\ M^2)\cdot(0.050\ \cancel{M})} = 9.0 \times 10^2\ \frac{1}{M^2 \cdot s}$$

Thus, the final rate equation is:

$$\underline{R = (9.0 \times 10^2\ \frac{1}{M^2 \cdot s}) \cdot [NO]^2[Cl_2]}$$

4. The rate equation will take on the form:

$$R = k[C_3H_6Br_2]^x[I^-]^y$$

To determine x and y, we look at trials where the concentration of one reactant stayed the same and the concentration of the other reactant changed. In trials 1 and 2, the concentration of I^- remained the same but the concentration of $C_3H_6Br_2$ doubled. When that happened, the rate doubled. This means that x = 1. In trials 1 and 3, the $C_3H_6Br_2$ concentration remained constant, but the I^- concentration doubled. When that happened, the rate doubled. This means y = 1. The rate equation, then, looks like:

$$R = k[C_3H_6Br_2][I^-]$$

To solve for k, we can use the data from any trial and plug it into our rate equation. We can then solve for k:

$$R = k[C_3H_6Br_2][I^-]$$

$$0.234\ \frac{M}{s} = k \cdot (0.100\ M) \cdot (0.100\ M)$$

$$k = \frac{0.234\ \frac{\cancel{M}}{s}}{(0.100\ \cancel{M}) \cdot (0.100\ M)} = 23.4\ \frac{1}{M \cdot s}$$

The overall rate equation, then is

$$\underline{R = (23.4\ \frac{1}{M \cdot s}) \cdot [C_3H_6Br_2][I^-]}$$

5. Since chemical reaction rate doubles for every 10 °C increment, then to increase the rate of the reaction by a factor of 8, I just need to raise the temperature by 3 ten degree increments. That way, I will multiply the old rate by 2x2x2, which equals 8. Thus, to increase the reaction rate by a factor if 8, I just raise the temperature by 30 degrees. Therefore, the new temperature should be 25 °C + 30°C = 55 °C.

6. A catalyst is used up in an early step and remade in a later step so that its concentration does not change. This is happening with Cl_2. It is a homogeneous catalyst, as it is in the same phase as the reactants.

7. It lowers the size of the hill.

SOLUTIONS TO THE EXTRA PRACTICE PROBLEMS FOR MODULE #15

1. Acidic solutions have pH levels under 7. Thus, solutions A and B are acidic.

2. The largest ionization constant corresponds to the strongest acid, which would produce the lowest pH. Thus, solution C is made from the acid with the largest ionization constant.

3. According to Equation (15.2), the equilibrium constant for this reaction is:

$$K = \frac{[NH_3]_{eq}[HCO_3^-]_{eq}}{[NH_4^+]_{eq}[CO_3^{2-}]_{eq}}$$

Plugging those equilibrium concentrations into the equation:

$$K = \frac{(1.2\ \cancel{M})(1.2\ \cancel{M})}{(0.80\ \cancel{M})\cdot(0.80\ \cancel{M})} = \underline{2.3}$$

4. The equation for the equilibrium constant here is:

$$K = \frac{[SO_3]_{eq}{}^2}{[SO_2]_{eq}{}^2[O_2]_{eq}}$$

If the concentrations are, in fact, equilibrium concentrations, then the equation should equal the value given for K.

$$K = \frac{(0.560\ \cancel{M})^2}{(0.280\cancel{M})^2(0.294\ M)} = 13.6\ \frac{1}{M}$$

If the reaction were at equilibrium, the value we just calculated would be equation to K. It is not, so the reaction is not at equilibrium. The calculated value is larger than K, so the value needs to decrease. This will happen if the reaction shifts towards the reactants.

5. a. Since this is an endothermic reaction, energy is a reactant. Raising the temperature, then, is like adding a reactant. Thus, the equilibrium will shift towards the products. This will increase the concentration of H_2.

b. When pressure is raised, the equilibrium shifts to the side with the fewest gas molecules. In this case, that's the reactants side. Thus, the concentration of NH_3 will increase.

c. If the concentration of NH_3 is lowered, the forward reaction rate will decrease, making the reverse reaction faster in comparison. This will cause a shift towards the reactants, making the concentration of N_2 decrease.

6. a. Since this is an exothermic reaction, energy is a product. Raising the temperature, then, is like adding a product. Thus, the equilibrium will shift towards the reactants. This will increase the concentration of CO.

b. When pressure is raised, the equilibrium shifts to the side with the fewest gas molecules. In this case, that's the products side. Thus, the concentration of CO will decrease.

c. If more carbon is added, nothing will happen, because changing the amount of solid will not stress the equilibrium.

d. If more CO is added, that will increase the rate of the forward reaction, making more products. Thus, the amount of carbon will increase. Its concentration will not change, but the amount will increase.

7. The ionization constant is simply the equilibrium constant for the acid ionization reaction. In order to determine the ionization reaction, you simply take the acid in its aqueous phase and remove an H^+. When you remove an H^+ from NH_4^+, you are left with NH_3. In the end, then, the aqueous acid is the reactant, and the H^+ and NH_3 (both in aqueous phase) will be the products:

$$NH_4^+ (aq) \leftrightharpoons H^+ (aq) + NH_3 (aq)$$

The equilibrium constant for this reaction is the ionization constant, K_a:

$$K_a = \frac{[H^+][NH_3]}{[NH_4^+]}$$

SOLUTIONS TO THE EXTRA PRACTICE PROBLEMS FOR MODULE #16

1. When a substance is made up of only one type of atom, the oxidation number of the atom is the charge of the substance divided by the number of atoms in the substance.

a. 0 b. +1 c. 0 d. 0

2. a. Rule #2 says that K will have an oxidation number of +1, and rule #6 says that O will have an oxidation number of -2. Since all of the oxidation numbers must add up to the overall charge, the Mn is +7. The oxidation numbers, then, are: K: +1, Mn: +7, O: -2

b. Rule #2 tells us that Na is +1 here. Rule #5 tells us that H is also +1 here. Rule number #6 tells us that O is -2. In order for the sum of all oxidation numbers to equal the overall charge (0), C must be +2. The oxidation numbers, then, are: Na: +1, C: +2, H: +1,O: -2

c. Rule #6 tells us that O is -2 here. Rule #5 tells us that H is +1 here. Since the sum of all oxidation numbers must equal the overall charge (0), Zn must be +2. The oxidation numbers, then, are: Zn: +2, O: -2, H: +1

d. This one where we use the last resort. We should always start with group 7A atoms when using the last resort, so Cl is -1. That makes S +2. The oxidation numbers, then, are: S: +2, Cl: -1

3. This atom gained three electrons, so it was reduced.

4. This atom lost four electrons, so it was oxidized.

5. a. This is not a redox reaction, because the oxidation numbers did not change.

b. This is a redox reaction, because Mg went from 0 to +2 and O went from 0 to -2.

c. This is not a redox reaction, because the oxidation numbers did not change.

d. This is a redox reaction, because H went from +1 to 0 and Cl went from -1 to 0.

6. In 5b, Mg was oxidized because it lost electrons, and O was reduced because it gained electrons.

In 5d, H was reduced because it gained electrons, and Cl was oxidized because it lost electrons.

7. In 5b, two electrons were transferred, because the charges changed by two.

In 5d, one electron was transferred, because the charges changed by one.

8. In this reaction, Al^{3+} is going from an oxidation number of +3 to an oxidation number of 0. This indicates that it is gaining electrons. Thus, the solution holding aqueous Al^{3+} will have electrons flowing into it. Electrons flow towards the Al^{3+}, so that container is positive (it attracts electrons) and thus will be the cathode. The Mg is going from an oxidation number of 0 to an oxidation number of +2. This means it loses electrons. Since it is losing electrons, the electrons are flowing away from the

container holding the solid Mg. This makes that container the negative side of the battery (it repels electrons), and it is thus the anode. The picture, then, looks like this:

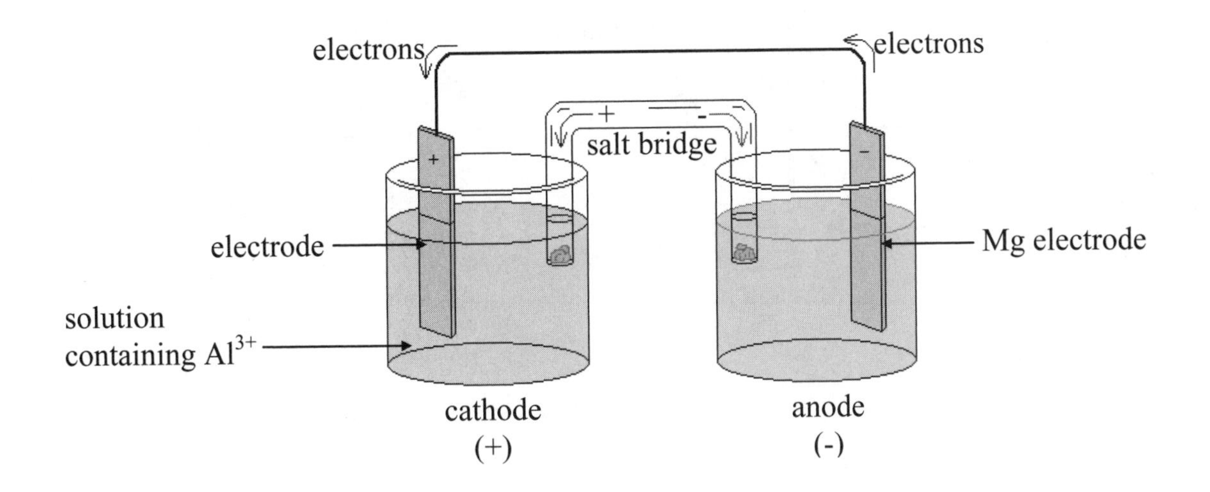

9. In this reaction, F_2 is going from an oxidation number of 0 to an oxidation number of -1. This indicates that it is gaining electrons. Thus, the solution holding aqueous F_2 will have electrons flowing into it. Electrons flow towards the F_2, so that container is positive (it attracts electrons) and thus will be the cathode. The Sn is going from an oxidation number of 0 to an oxidation number of +2. This means it loses electrons. Since it is losing electrons, the electrons are flowing away from the container holding the solid Sn. This makes that container the negative side of the battery (it repels electrons), and it is thus the anode. The picture, then, looks like this:

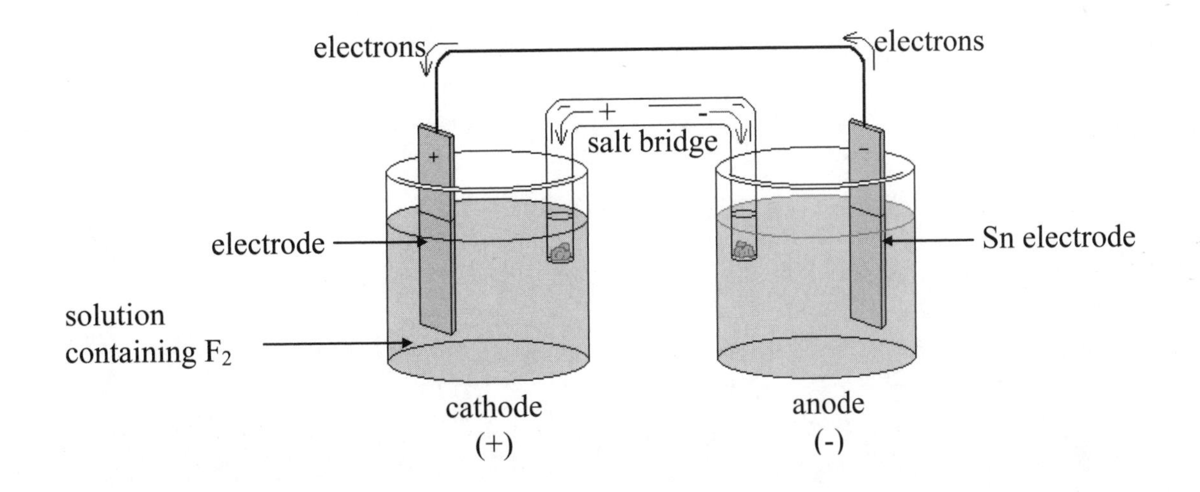

10. In this reaction, Cu^{2+} is going from an oxidation number of +2 to an oxidation number of 0. This indicates that it is gaining electrons. Thus, the solution holding aqueous Cu^{2+} will have electrons flowing into it. Electrons flow towards the Cu^{2+}, so that container is positive (it attracts electrons) and thus will be the cathode. The Cr is going from an oxidation number of 0 to an oxidation number of +3. This means it loses electrons. Since it is losing electrons, the electrons are flowing away from the container holding the solid Cr. This makes that container the negative side of the battery (it repels electrons), and it is thus the anode. The picture, then, looks like this:

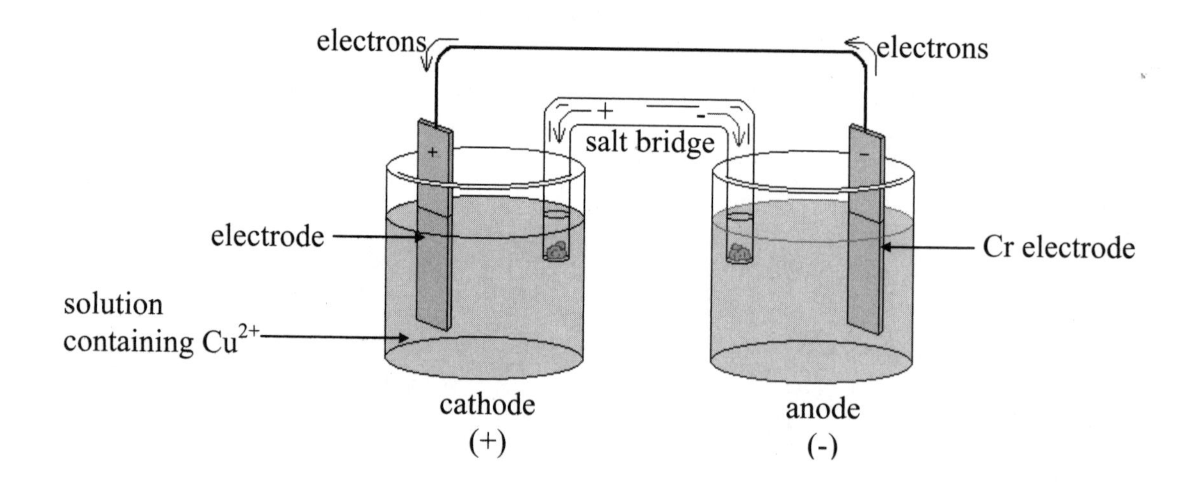
electrons
electrons
+
-
salt bridge
+
-
electrode
Cr electrode
solution
containing Cu2+
cathode
(+)
anode
(-)

TEST FOR MODULE #1

Be sure to write all of your answers down with the proper number of significant figures and to list the units that go with your answers!

1. What are the meanings for the metric prefixes "milli," "centi," and "kilo?"

2. Which is heavier, a 0.3 g rock or a 30.0 mg rock?

3. The following numbers are the results of several measurements of a football field (which is supposed to be 100.0 yards long):

    a. 113.1 yards
    b. 1.0×10^2 yards
    c. 99.126 yards

Which of these three numbers represents the most precise measurement? Which is the most accurate?

4. If we observe a fishing bobber float on the surface of the water, what can we conclude about the density of the bobber compared to the density of the water?

5. What is the volume of the liquid in the following graduated cylinder?

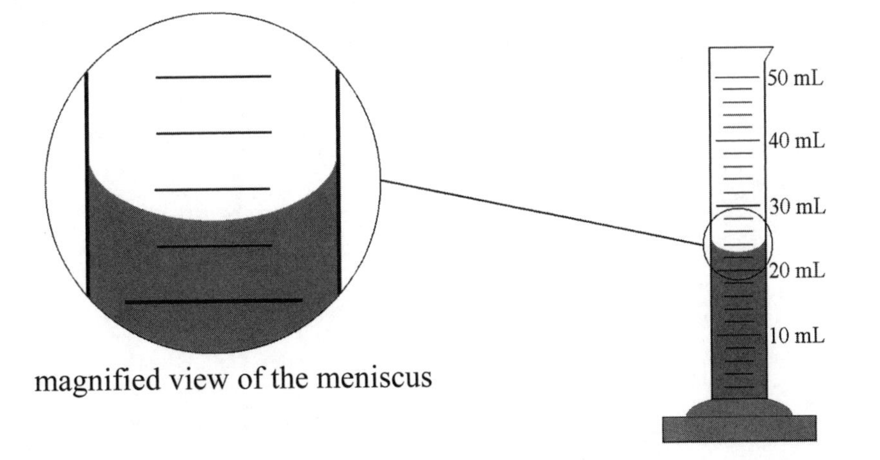

magnified view of the meniscus

6. If the mass of the liquid in problem #5 was 32.13 grams, what would its density be?

7. How many cm are in 16.2 m?

8. If an object has a mass of 345.6 mg, what is its mass in grams?

9. If a football field is 100.0 yards long, how many miles long is it? (1 yard = exactly 3 feet, 1 mile = 5.280×10^3 feet).

10. The density of gold is 19.3 grams per mL. A miner finds a gold-colored nugget whose volume is 34.2 mL and whose mass is 661 grams. Is it really a nugget of gold?

11. A fisherman wants to determine the volume of his lead sinker. If lead's density is 11.4 grams per cc and the sinker has a mass of 0.123 kg, what is the volume of the sinker?

12. Convert the number 3478 to scientific notation.

13. Convert the number 1.245×10^{-4} into decimal notation.

14. The volume of a sphere is given by the equation

$$V = \frac{4}{3}\pi r^3$$

Where $\pi = 3.1416$, and r is the radius of the sphere. If a sphere's radius is 3.1 m, what is its volume in liters?

15. The mass of an object is measured on earth and also on the moon. How do the two measured masses compare?

TEST FOR MODULE #2

1. Why does a person do no work if he or she pushes against a car that is too heavy to move?

2. Write the first law of thermodynamics in your own words. Use it to explain where the flame of a burning candle gets its heat.

3. Explain the difference between kinetic and potential energy. Give at least one example of each.

4. What is the relationship between a food Calorie and a chemistry calorie?

5. If the temperature of an object starts to increase, is the object gaining or losing energy?

6. What two units do we use for measuring energy? What is the relationship between them?

7. Convert 100.0 K to both Celsius and Fahrenheit.

8. The specific heat of plastic is fifty times larger than the specific heat of lead and ten times larger than the specific heat of stone. Equal masses of lead, plastic, and stone have the same initial temperature. They are each given same amount of energy. Which ends up the hottest?

9. Many recipes call for cooking food at a very constant temperature of 212 °F (100 °C). Why do these recipes recommend that you boil the food to achieve this constant temperature?

10. A 10.0 kg copper skillet ($c = 0.3851 \frac{J}{g \cdot {}^{\circ}C}$) must be heated from room temperature (25.0 °C) to a temperature of 175.0 °C. How many Joules of heat are required?

11. A 1.50 kg iron object ($c = 0.4521 \frac{J}{g \cdot {}^{\circ}C}$) at 110.0 °C releases 5,505 Joules of heat. What is its new temperature?

12. If a 50.0 g object needs 1,145 Joules to increase its temperature by 10.0 °C, what is its specific heat capacity?

13. What are the basic components of a calorimeter? Briefly describe how a calorimetry experiment is done and what you can learn from it.

14. In a calorimetry experiment, a calorimeter is filled with 125 grams of water. The initial temperature of the water is 22.3 °C. A 50.0 g chunk of metal at 123.0 °C is dropped into the calorimeter, and the temperature increases to 24.5 °C. What is the specific heat of the metal? You can ignore the heat absorbed by the calorimeter.

(There is one more problem on the next page.)

15. In a calorimetry experiment, an unknown mass of glass (c = 0.8372 $\frac{J}{g \cdot {}^{o}C}$) at 90.0 °C is dropped into a calorimeter (c = 1.23 $\frac{J}{g \cdot {}^{o}C}$, m = 7.0 g) that contains 75.0 g of water. If the temperature increases from 24.3 °C at the beginning of the experiment to 26.6 °C by the end, what was the mass of the glass?

TEST FOR MODULE #3

1. If a substance can be decomposed, is it an element or a compound?

2. If a substance can conduct electricity when dissolved in water, what do we classify it as?

3. State the law of definite proportions in your own words.

4. A chemist does two experiments. In one experiment, he finds that 14.0 g of nitrogen combine with 16.0 g of oxygen to make 30.0 g of a compound he calls "compound a." In another experiment, the chemist finds that 48.0 g of oxygen combine with 42.0 g of nitrogen to make 90.0 g of a compound he calls "compound b." The chemist states that the law of definite proportions tells him that "compound a" and "compound b" are two completely different compounds. Why is he wrong?

5. A substance is seen to be brittle and not conduct electricity. Is it most likely a metal or a nonmetal?

6. Which theory of matter do scientists believe today: the discontinuous theory of matter or the continuous theory of matter?

7. All scientists today believe that atoms do, in fact, exist. Has their existence been proven?

8. A chemist decomposes 100.1 grams of a substance into 12.1 grams of carbon and 40.0 g of magnesium. The only other product of the decomposition was oxygen, but the gas escaped from the chemist's experimental setup. What was the mass of oxygen gas that escaped?

9. In an experiment to determine how to make sulfur trioxide, a chemist combines 32.0 g of sulfur with 50.0 g of oxygen. She finds that she made 80.0 g of sulfur trioxide and had 2.0 g of left over oxygen. How would the chemist make 100.0 g of sulfur trioxide so that she has no leftovers?

10. To make 44.0 grams of carbon dioxide, you must combine 12.0 g of carbon with 32.0 g of oxygen. If a chemist combines 120.0 g of carbon with 160.0 g of oxygen, how many grams of carbon dioxide will be made? If a substance is left over, indicate whether it is carbon or oxygen, and also determine how many grams are left over.

11. You are told that a molecule contains a total of 17 atoms. If the formula is $C_5H_{10}Cl_x$, what must "x" be?

12. Which of the following atoms would be considered nonmetals?

W, Ne, Xe, Mn, Fe, S

13. Which of the following compounds would conduct electricity when dissolved in water?

C_3H_6ClF, $NaNO_3$, CS_2, $BaSO_4$

14. Name the following compounds:

a. Al_2S_3 b. SF_2 c. C_2F_6 d. BeF_2

15. What is the chemical formula of sulfur trioxide?

TEST FOR MODULE #4

1. What are the four classifications that we learned for matter?

2. What are the two classifications for the changes that matter undergoes?

3. If a substance is an element, is it a pure substance or a mixture?

4. Classify the following as element, compound, homogeneous mixture, or heterogeneous mixture:

   a. Fruit salad   b. Aluminum foil   c. A pile of $CaCO_3$ d. Fertilizer dissolved in water

5. What are the differences between solid HCl and liquid HCl?

6. If a substance goes from the gas phase to the liquid phase, was it heated or cooled?

7. Ice floats in water. Does frozen HNO_3 float in liquid HNO_3?

8. Classify the following changes as physical or chemical:

   a. A lawn gets mowed.
   b. A frozen turkey thaws.
   c. Wood burns in a fireplace.

9. Which of the following elements are homonuclear diatomics? Write their chemical formulas.

   carbon, nitrogen, fluorine, sodium

10. Is the following equation balanced?

$$2HBr\ (aq) + Ca\ (s) \rightarrow CaBr_2\ (aq) + H_2\ (g)$$

11. Balance the following equation:

$$H_2O\ (l) + Na\ (s) \rightarrow NaOH\ (aq) + H_2\ (g)$$

12. Under the right conditions, chlorine gas will react with gaseous ethylyne (C_2H_2) to make liquid quadrachloroethane ($C_2H_2Cl_4$). Write the balanced chemical equation.

13. When C_9H_{20} reacts with oxygen, it makes carbon dioxide and water. Write the balanced chemical equation.

14. Balance the following equation for the combustion of ethanol:

$$C_2H_6O\ (l) + O_2\ (g) \rightarrow CO_2\ (g) + H_2O\ (g)$$

15. How does a balanced chemical equation ensure that mass is conserved during a chemical reaction?

TEST FOR MODULE #5

(1.00 amu = 1.66×10^{-24} g)

For the following 4 questions, classify the reaction as formation, decomposition, combustion, or none of these.

1. $2H_2CO_3\ (aq) \rightarrow 2C\ (s) + 3O_2\ (g) + 2H_2\ (g)$

2. $2AgCl\ (aq) + Ca\ (s) \rightarrow CaCl_2\ (aq) + 2Ag\ (s)$

3. $2C_2H_2 + 5O_2 \rightarrow 4CO_2 + 2H_2O$

4. $Br_2\ (g) + C\ (s) + H_2\ (g) \rightarrow CH_2Br_2\ (l)$

5. Write the chemical equation for the formation of PH_2Br.

6. Write the chemical equation for the decomposition of Na_2CrO_4.

7. Write the chemical equation for the complete combustion of liquid $C_{10}H_{22}$. Include phase symbols in your answer.

8. What is the mass of a Pr atom, in kg?

9. What is the mass of an $HClO_4$ molecule in amu?

10. How many moles are in 34.5 g of $NaBrO_3$?

11. If a chemist has 12.3 moles of $N_2H_4O_3$, what is the mass of the sample?

12. If a chemist has 0.45 moles of K_2CO_3 and causes it to decompose into its constituent elements, how many moles of oxygen will form?

13. Which sample has more atoms in it: 150.0 g of gold (Au), or 10.0 g of lithium?

14. If a fireplace or furnace has poor airflow, what should you be concerned about?

15. A chemist has one mole of an unknown molecule. The mass of this sample is 111.1 grams. After doing some tests, the chemist determines that the molecule has 2 chlorine atoms in it and one other atom she cannot identify. In other words, the molecule's formula is XCl_2, where X is unknown. Based on the mass of the sample, what is atom X?

TEST FOR MODULE #6

1. What is a stoichiometric coefficient?

2. Fill in the blanks:

   In the equation

$$2HF + CaO_2H_2 \rightarrow 2H_2O + CaF_2$$

   2 moles of HF react with ___ mole(s) of CaO_2H_2 to make ____ moles(s) of H_2O and ____ mole(s) of CaF_2.

3. The chemical reaction in problem #2 is run with HF as the limiting reactant. What would happen to the amount of H_2O produced in the reaction if the number of moles of CaO_2H_2 were doubled but the number of mole of HF were kept the same?

4. State Gay-Lussac's Law and its limitations.

5. An unknown compound is decomposed into 6.9 g of Na, 3.1 g of P and 6.4 g of oxygen. What is its empirical formula?

6. One type of stomach antacid, $Al(OH)_3$, can be made in the following way:

$$Al_2(SO_4)_3 + 6NaOH \rightarrow 2Al(OH)_3 + 3Na_2SO_4$$

How many moles of $Al(OH)_3$ can be made with 2.3 moles of NaOH and excess $Al_2(SO_4)_3$?

7. AgBr, a chemical used in photography, can be made with this reaction:

$$2AgNO_3 + CaBr_2 \rightarrow 2AgBr + CaN_2O_6$$

$AgNO_3$ is very expensive, so in making AgBr, you want to be sure this reactant is all used up. If a chemist starts with 15.0 g of $AgNO_3$, how many grams of $CaBr_2$ must be added to use up all of the $AgNO_3$?

8. Is Al_2Cl_6 an empirical formula? If not, convert it to an empirical formula.

9. A chemist performs the following reaction:

$$6ClO_2\ (g) + 3H_2O\ (l) \rightarrow 5HClO_3\ (l) + HCl\ (g)$$

If the chemist wants to make 50.0 g of $HClO_3$, what is the minimum number of grams of ClO_2 that she can use?

10. If the reaction in problem #9 was run with excess water, how many liters of HCl would be produced from 10.0 liters of ClO_2?

11. Why is the following problem unsolvable?

If the reaction in problem #9 was run with excess water, how many liters of $HClO_3$ would be produced from 10 liters of ClO_2?

12. A compound has an empirical formula of CH_2 and a molecular mass of 70.1 amu. What is its molecular formula?

13. What is wrong with the following statement?

In the reaction:

$$N_2 + 3H_2 \rightarrow 2NH_3$$

1 gram of N_2 reacts with 3 grams of H_2 to make 2 grams of NH_3

14. Consider the following reaction:

$$Fe_2O_3\ (s)\ +\ 3CO\ (g) \rightarrow 2Fe\ (s)\ +3\ CO_2\ (g)$$

If 2 moles of Fe_2O_3 are reacted with 3 moles CO, what is the limiting reactant?

15. With the reaction and quantities given in problem 14, how many moles of Fe would be formed?

TEST FOR MODULE #7

($c = 3.0 \times 10^8$ m/s, "nano" = 10^{-9}, $h = 6.63 \times 10^{-34}$ J/Hz)

1. Two masses are positively charged. Will they repel or attract each other?

2. A substance has 34 positive charges and 34 negative charges. What is its overall charge?

3. What was Crooke's contribution to the science of atomic structure?

4. How many protons, neutrons, and electrons are in a ^{141}Ba atom?

5. What is the symbol for an atom with 45 protons, 57 neutrons, and 45 electrons?

6. What is the definition of isotopes?

7. You have two light bulbs of equal brightness. One is yellow and one is indigo. What can you say about their amplitudes? Which emits light of higher frequency?

8. What is the frequency of light that has a wavelength of 10.0 nm?

9. If the energy of light emitted from an atom is 2.3×10^{-15} Joules, what is its wavelength?

10. Which is the heaviest: a neutron, a proton, or an electron?

11. An atom is emitting light. What are its electrons doing?

12. Order the three orbital shapes we have studied in terms of increasing energy.

13. What is wrong with the following electron configuration?

$$1s^2 2s^3 2p^6 3s^2 3p^6 4s^2 3d^5$$

14. Give the full electron configurations for the following atoms:

    a. Ca b. Nb c. Fe

15. Give abbreviated electron configurations for the following atoms:

    a. Fr b. Cl c. Cd

TEST FOR MODULE #8

1. Which of the following atoms will have similar chemistry?

Si, P, S, N, As

2. How many valence electrons are in this electron configuration?

$$1s^2 2s^2 2p^6 3s^2 3p^6 4s^2 3d^{10} 4p^2$$

3. What columns in the periodic chart represent the transition metals?

4. What column in the periodic chart contains the atoms with an ideal electron configuration?

5. What type of charge do metal ions develop in ionic compounds?

6. When atoms share electrons to make a molecule, is the compound ionic or covalent?

7. When trying to determine whether or not an atom gives up its electrons easily, do chemists look at electronegativity or ionization potential?

8. What does ozone do to make life on earth possible?

9. What is the Lewis structure for Al?

10. Write the chemical formulas for the following compounds:

    a. potassium arsenide (arsenic is abbreviated As)
    b. iron(III) oxide (Iron's symbol is Fe)
    c. calcium sulfide

11. Which of the following atoms gives up its electrons most easily?

Mg, Na, P

12. Order the following atoms in terms of increasing electronegativity:

Ba, Mg, Ca, Sr.

13. Draw the Lewis structures for the following molecules:

a. PH_3 b. Br_2 c. SiH_4 d. OCl_2 e. SiO_2 f. CS_2 g. FSiN

TEST FOR MODULE #9

1. How many oxygen atoms are in an $Al_2(CO_3)_3$ molecule?

2. Polyatomic ions are often said to be a cross between covalent compounds and ionic compounds. Why can they be characterized in such a way?

3. Give the chemical formulas for the following molecules:

a. calcium acetate　　b. barium chromate　　c. ammonium carbonate

4. Give the balanced chemical equation for the reaction that occurs when aqueous ammonium chloride reacts with aqueous barium nitrate to produce solid barium chloride and aqueous ammonium nitrate.

5. In VSEPR theory, which repels other electron groups more strongly, a bond or a non-bonding electron pair?

6. Explain, in your own words, why molecules sometimes attain three-dimensional shapes rather than just being flat.

7. Determine the shape of an I_2 molecule. Give its bond angles and draw a picture of it.

8. Determine the shape of a NHF_2 molecule. Give its bond angles and draw a picture of it.

9. Determine the shape of a CHI_2Cl molecule. Give its bond angles and draw a picture of it.

10. Determine the shape of an I_2O molecule. Give its bond angles and draw a picture of it.

11. Determine the shape of an SiS_2 molecule. Give its bond angles and draw a picture of it.

12. Classify the following molecules as ionic, polar covalent, or purely covalent:

a. NHF_2　　b. CHI_2Cl　　c. I_2O　　d. SiS_2　　e. KNO_3　　f. CF_4　　g. I_2

13. Which of the molecules in problem #12 should be able to dissolve in water?

TEST FOR MODULE #10

1. A substance tastes sour and causes blue litmus paper to turn red. Is it an acid or a base?

2. In the following equations, identify the acid and the base:

   a. $C_2H_6O + CO_3^{2-} \rightarrow HCO_3^{1-} + C_2H_5O^{1-}$
   b. $HNO_3 + H_2O \rightarrow H_3O^{1+} + NO_3^{1-}$
   c. $HCl + KOH \rightarrow KCl + H_2O$

3. What is the chemical definition of a base?

4. Which of the following acids is (are) diprotic?

   a. HNO_3 b. H_2CO_3 c. H_3PO_4 d. HSO_4^{1-}

5. What is an amphiprotic substance? Give an example of one.

6. You have a brightly colored substance that you think might be an indicator. How would you make sure that it is one?

7. What is wrong with the following statement:

"In a titration, the endpoint occurs when the number of moles of acid added equals the number of moles of base in the original solution, or vice versa"

8. Give a balanced chemical equation for the reaction between phosphoric acid (H_3PO_4) and CsOH.

9. What is the balanced chemical equation that represents the reaction of HCl with $Be(OH)_2$?

10. Write the balanced chemical equation for the reaction of HF with water.

11. What is the concentration (in M) of an ammonia (NH_3) solution if 12.23 grams of ammonia are dissolved in enough water to make 560.0 mL of solution?

12. If a chemist has a stock solution of HBr that is 10.0 M and would like to make 450.0 mL of 3.0 M HBr, how would he or she do it?

13. A chemist reacts 30.0 mL of 5.6 M HCl with an excess of $Mg(OH)_2$. How many grams of magnesium chloride will be produced?

14. If a chemist titrates 120.0 mL of NaOH with a 5.0 M solution of HCl and requires 56.0 mL of the acid to reach the endpoint, what is the concentration of the NaOH?

15. If a chemist titrates 300.0 mL of H_2SO_4 with a 3.0 M solution of NaOH and requires only 3.4 mL of the base to reach the endpoint, what is the concentration of the sulfuric acid?

TEST FOR MODULE #11

1. Define solubility.

2. If a solute's solubility in a liquid solvent is not affected by temperature or pressure, is the solute most likely a solid, liquid, or gas?

3. If a solute's solubility in a liquid solvent decreases when the temperature increases, is the solute most likely a solid, liquid, or gas??

4. A chemist is trying to dissolve a solid in water. If the chemist feels the beaker get hot while trying to make the solution, does the solute dissolve exothermically or endothermically?

5. A chemist finds out that a sample of alcohol has been contaminated with a gas. What can the chemist do to get rid of the majority of this dissolved contaminant?

6. What kind of solute (solid, liquid, or gas) usually increases in solubility with increasing temperature?

7. Pick the solute that will raise a liquid's boiling point the most, assuming that the concentration of each solute is the same:

$$NH_3, Na_3PO_4, K_2CO_3, NaCl$$

8. A student makes a solution by measuring out 100 grams of solute and dissolving it in lots of water. The student then measures the mass of the resulting solution. He converts the mass of the solute into moles and divides that by the mass (in kg) of the solution. He reports the result as the molality of the solution. Why is the student wrong?

9. When NaI is added to a solution of $Cu(NO_3)_2$, the following reaction occurs:

$$2NaI\ (aq) + Cu(NO_3)_2\ (aq) \rightarrow CuI_2\ (s) + 2NaNO_3\ (aq)$$

If a chemist needs to make 350.0 g of CuI_2, how many mL of a 3.3 M solution of NaI must be used? Assume that there is excess copper (II) nitrate.

10. Hydrogen peroxide is used to clean small wounds. It has a limited shelf-life, however, because the following decomposition reaction occurs:

$$2H_2O_2\ (aq) \rightarrow 2H_2O\ (l) + O_2\ (g)$$

This is why you often hear a gas release when you open a bottle of old hydrogen peroxide. If 35 mL of a 1.20 M solution of hydrogen peroxide completely decomposes, how many grams of oxygen are produced?

11. What is the molality of a solution made from 5.61 moles of KCl and 2.11 kg of water?

12. If you have 670.0 grams of water and wish to make a 2.13 m solution of KBr, how many grams of the solute would you have to add to the water that you have?

13. When a certain amount of MgF_2 is added to water, the freezing point lowers by 3.5 °C. What was the molality of the magnesium fluoride? (K_f for water = $1.86 \frac{^\circ C}{m}$)

14. Benzene (a popular purely covalent solvent) has a freezing point of 5.5 °C and a K_f of $5.12 \frac{^\circ C}{m}$. If 30.0 g of CCl_4 are dissolved in 350.0 g of benzene, what will the freezing point of the solution be?

15. Acetic acid has a boiling point of 118.5 °C and a K_b of $3.08 \frac{^\circ C}{m}$. What is the boiling point of a 3.20 m solution of $Ca(NO_3)_2$ in acetic acid?

TEST FOR MODULE #12

1. State Charles's Law.

2. Which of the following is not a pressure unit?

atm, Pa, kPa, torr, pound, mmHg

3. Why is the Kelvin temperature scale called an absolute temperature scale?

4. Under which of the following conditions would you expect gases to behave ideally?

    a. T = 100 K, P = 3.4 atm
    b. T = 300 K, P = 0.9 atm
    c. T = 150 K, P = 1.0 atm
    d. T = 273K, P = 50.5 atm

5. What is the true definition of boiling point?

6. An experiment is run in which hydrogen gas is collected in a vessel over water. The first experiment is run at a temperature of 90 °C and the second is run at 15 °C. Everything except the temperature is exactly the same in the two experiments. In which experiment will the mole fraction of the hydrogen in the vessel be greatest?

7. What big mistake did Darwin make when he formulated his evolutionary theory?

8. A hot air balloon is filled with 1.89×10^2 liters of air at 21 °C. If atmospheric pressure does not change, how hot must the air become in order to increase the volume to 4.5×10^2 liters?

9. A weather balloon is filled with 34.6 liters of helium at 24 °C and 1.02 atm. As the weather balloon rises, the surrounding temperature decreases, as does the pressure. When the balloon reaches an altitude where the temperature is 11 °C and the pressure is 0.21 atms, what will the volume of the balloon be?

10. The vapor pressure of water is 30.0 torr at 29 °C. If a reaction at that temperature produces oxygen which is collected over water, what total pressure will have to be in the reaction vessel to ensure that the oxygen collected had a partial pressure of 567 torr?

11. The gas mixture for an acetylene torch should be about 12.0 grams of oxygen for every 100.0 grams of acetylene (C_2H_2). What is the mole fraction of each component in such a mixture?

12. What is the mole fraction of each component of a gas mixture that contains sulfur trioxide at a pressure of 1.45 atm and sulfur dioxide at a pressure of 0.32 atms?

13. If 345.1 grams of CO_2 are placed in a vessel whose volume is 32.1 liters at a temperature of 20.0 °C, what will the pressure be?

14. **Iron rusts** according to the following reaction:

$$4Fe\ (s) + 3O_2\ (g) + 4H_2O\ (l) \rightarrow 2Fe_2O_3(H_2O)_2$$

If 1.4 liters of oxygen were added to excess water and iron at a pressure of 0.97 atms and a temperature of 55 °C, how many grams of iron would rust?

15. The bright light that comes from a flash bulb is due to a quick combustion of magnesium, according to the following chemical equation:

$$2Mg\ (s) + O_2(g) \rightarrow 2MgO\ (s)$$

If a flashbulb contains 1.3 g of Mg, what is the minimum pressure of O_2 needed to burn up that magnesium? Assume the flashbulb has a volume of 10.0 mL and a temperature of 25 °C.

TEST FOR MODULE #13

(Use Tables 13.1 - 13.4 for this test.)

1. What kinds of reactions feel hot to the touch, exothermic or endothermic?

2. State Hess's Law and write down the equation that results from it.

Questions 3 - 5 refer to the following energy diagrams:

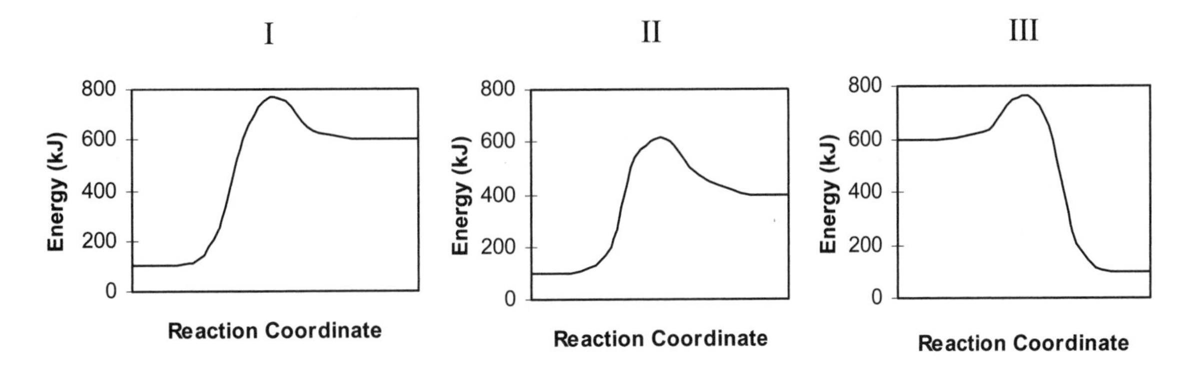

3. Which diagram(s) represent(s) an endothermic reaction?

4. Which diagram represents the reaction that is easiest to start?

5. What is the ΔH for the reaction represented by diagram II?

6. If I have 30.0 grams of carbon in one pile and 50.0 grams of carbon in the other, which sample of carbon has more entropy?

7. If a reaction is endothermic and has a positive ΔS, what should you do to the temperature in order to make sure the reaction will run?

8. Use bond energies to determine the ΔH for the combustion of methane (CH_4).

9. Use Hess's Law to determine the ΔH for the combustion of liquid methanol (CH_3OH).

10. Using your answer from #9, how much energy will you get from burning 245.0 g of methanol in excess oxygen?

11. If you were to calculate a reaction's ΔH using bond energies, you would most likely get a slightly different answer than when you use Hess's Law. Why?

12. Give the sign of ΔS in the following reactions:

    a. $CaSO_4\ (s) \rightarrow CaO\ (s) + SO_3\ (g)$

    b. $2H_2\ (g) + O_2\ (g) \rightarrow 2H_2O\ (g)$

    c. $2AgNO_3\ (aq) + Mg(OH)_2\ (aq) \rightarrow 2AgOH\ (s) + Mg(NO_3)_2\ (aq)$

13. The ΔS for the following reaction:

$$C_2H_4\ (g) + 3O_2\ (g) \rightarrow 2CO_2\ (g) + 2H_2O\ (g)$$

is 209.6 J/K. What is the absolute entropy of O_2 (g)?

14. What is ΔG for the following reaction at 298 K?

$$2C_6H_6\ (l) + 15O_2\ (g) \rightarrow 12CO_2\ (g) + 6H_2O\ (g)$$

15. If a reaction has a ΔH of -120.0 kJ/mole and is spontaneous for all temperatures less than 200.0 K, what is its ΔS?

TEST FOR MODULE #14

1. A chemist makes a solution of NaI (aq) to use as a reactant in an experiment. She makes 500 mL of the solution and only uses 100 mL in the experiment. When she does the experiment, she measures the rate of the reaction that involves NaI. She then takes the 400 mL of left over NaI (aq) and boils it down to only 100 mL. She then allows the NaI solution to cool back down to its original temperature and runs the same experiment again, with this new sample of NaI (aq). Will the rate she measures be larger, smaller, or the same as the rate she measured originally?

2. Explain in your own words why increasing the temperature of a reaction will increase its rate.

3. If the order of a chemical reaction with respect to one of its reactants is zero, how does that reactant's concentration affect the rate of the chemical reaction?

4. What part of the rate equation depends on the temperature of the reaction?

5. Two different reactions have exactly the same activation energy. What can you say about their reaction rates? Assume that temperature, reactant concentration, and order are the same in both reactions.

6. Give the definitions for heterogeneous and homogeneous catalysts.

7. A student is studying the rate of the following chemical reaction:

$$C_2H_4O + NaOH \rightarrow H_2O + NaC_2H_3O$$

Knowing that this is an exothermic reaction, he is measuring the rate of the reaction by timing how quickly the reaction vessel heats up. He notices that if he adds HCl to this reaction, the rate increases dramatically. He also determines that the HCl is being used up during the reaction. Is the HCl a catalyst for this reaction?

8. A chemist does a rate analysis on the following reaction:

$$C_3H_6Cl_2 + 3Br^- \rightarrow C_3H_6 + 2Cl^- + 3Br^-$$

She gets the following data:

Trial	Initial Concentration of $C_3H_6Cl_2$ (M)	Initial Concentration of Br^- (M)	Instantaneous Reaction Rate (M/s)
1	0.080	0.080	0.116
2	0.160	0.080	0.232
3	0.160	0.160	0.232

What is the rate equation for this reaction?

9. A chemist does a rate analysis on the following reaction:

$$2NO + I_2 \rightarrow 2NOI$$

He gets the following data:

Trial	Initial Concentration of NO (M)	Initial Concentration of I_2 (M)	Instantaneous Reaction Rate (M/s)
1	1.5	1.0	1.25
2	1.5	2.0	2.50
3	3.0	2.0	10.0

What is the rate equation for this reaction?

10. Experiment tells us that the following reaction:

$$CH_3OH + KF \rightarrow KOH + CH_3F$$

is first order with respect to each of its reactants. We also know that the rate constant for this reaction is equal to 0.0123 1/Ms at 10 °C. What is the instantaneous rate for this reaction at 10 °C if the initial concentration of CH_3OH is 0.59 M and the initial concentration of KF is 1.22 M?

11. A certain chemical reaction has three reactants. According to experiment, it is first order with respect to one of the reactants and second order with respect to each of the other 2 reactants. What is the overall order of the reaction?

12. A chemist runs a reaction at 50 °C and determines its rate to be 0.00451 M/s. If she increases the temperature to 80 °C, what will the rate of the reaction be?

13. A chemist runs a reaction at 25 °C and determines that the reaction proceeds too quickly. He decides that the reaction needs to be slowed down by a factor of 4. At what temperature should he run the reaction?

14. The rate of an exothermic reaction is studied. First, the rate is measured for the reaction with no catalyst. Second, the rate of the reaction is measured when catalyst A is used, and it is found that the rate with catalyst A is three times higher than the rate without a catalyst. Finally, the rate is measured when catalyst B is used. The rate of the reaction when catalyst B is used is ten times higher than the rate of the reaction when no catalyst is used. Draw three energy diagrams for this reaction. The first should represent the reaction with no catalyst. The second should represent the reaction with catalyst A. The final energy diagram should represent the reaction with catalyst B. Do not worry about labeling the y-axis of the graph, as you do not know the energies involved.

15. Identify the catalyst in the following mechanism:

Step 1. $Ce^{4+} + Mn^{2+} \rightarrow Ce^{3+} + Mn^{3+}$
Step 2. $Ce^{4+} + Mn^{3+} \rightarrow Ce^{3+} + Mn^{4+}$
Step 3. $Mn^{4+} + Tl^{+} \rightarrow Tl^{3+} + Mn^{2+}$

TEST FOR MODULE #15

1. Based on the following graph, at approximately what time did this reaction reach equilibrium?

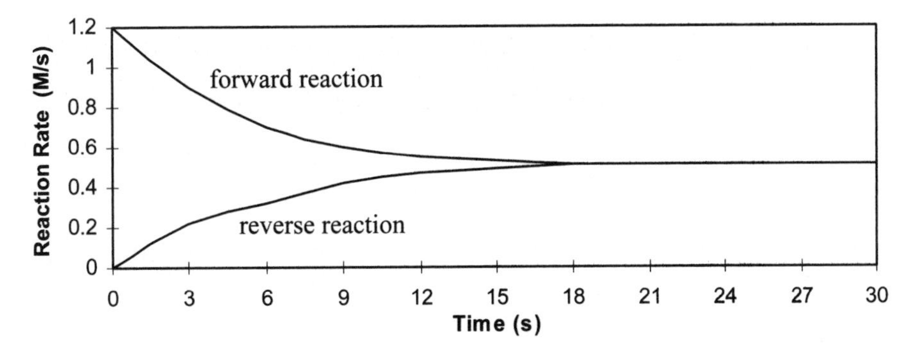

2. What does Le Chatelier's principle say ?

3. What does the pH scale measure?

4. What is the acid ionization reaction for HNO_3?

5. Based on the following graph, at approximately what time was the equilibrium stressed?

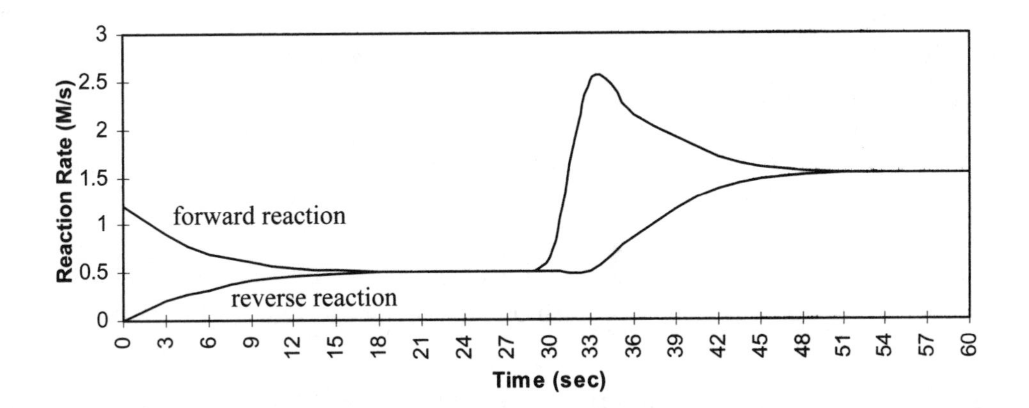

6. Five solutions of equal concentration have the following pH:

Solution A: pH = 7.0
Solution B: pH = 14.0
Solution C: pH = 9.1
Solution D: pH = 5.5
Solution E: pH = 1.2

Which solution contains the strongest acid? Which contains the weakest acid? Which contains the strongest base? Which contains the weakest base?

7. Five acids have the following pH in equal concentrations:

Acid A: pH = 6.8
Acid B: pH = 5.0
Acid C: pH = 1.1
Acid D: pH = 4.5
Acid E: pH = 2.1

Order the acids in terms of increasing ionization constants.

8. The following reaction:

$$2NOBr\ (g) \leftrightharpoons 2NO\ (g) + Br_2\ (g)$$

reached equilibrium when the concentrations were: [NOBr] = 0.10 M, [NO] = 0.010 M, and $[Br_2]$ = 0.0050 M. What is the equilibrium constant for this reaction?

9. A chemist knows that the equilibrium constant for the following reaction:

$$BaSO_3\ (s) \leftrightharpoons BaO\ (s) + SO_2\ (g)$$

is 0.345 M. What is the concentration of SO_2 when the reaction reaches equilibrium?

10. Three chemical equilibria are written below, along with their equilibrium constants. If any of these equations can be written with a single arrow, do so.

a. $2NO_2\ (g) + Cl_2\ (g) \leftrightharpoons 2NOCl\ (g)$ $K = 3.7 \times 10^4\ 1/M$
b. $2SO_2\ (g) + O_2\ (g) \leftrightharpoons 2SO_3\ (g)$ $K = 0.63\ 1/M$
c. $N_2O_4 \leftrightharpoons 2NO_2$ $K = 1.1 \times 10^{-5}\ M$

11. A chemist is studying the following reaction:

$$SO_2\ (g) + H_2O\ (l) \leftrightharpoons H_2SO_3\ (aq) \qquad K = 2.1 \times 10^{-3}$$

She measures the concentrations of H_2SO_3 and SO_2 and their concentrations are 0.23 M and 0.35 M, respectively. Is the reaction at equilibrium? If not, which way must the reaction shift to attain equilibrium?

12. Consider the following reaction that has reached equilibrium:

$$MgSO_3\ (s) + H_2O\ (l) \leftrightharpoons H_2SO_3\ (aq) + MgO\ (aq)$$

a. What will happen to the concentration of H_2SO_3 if extra $MgSO_3$ is added?
b. What will happen to the amount of MgO if H_2SO_3 is removed?
c. What will happen to the concentration of MgO if more H_2O is added?

13. Consider the following reaction that has reached equilibrium:

$$N_2\ (g) + 2O_2\ (g) \leftrightharpoons 2NO_2\ (g) \qquad \Delta H = 52.2\ kJ$$

a. What will happen to the concentration of O_2 if the temperature is raised?
b. What will happen to the concentration of NO_2 if the pressure is raised?
c. What will happen to the concentration of N_2 if the pressure is lowered?

14. What is the equation for the acid ionization constant of $HClO_2$?

15. What is the equation for the base ionization constant of CH_5P?

TEST FOR MODULE #16

1. Explain why oxidation numbers are not actually real charges.

2. An atom has an oxidation number of -3 in one molecule. If it is placed in another molecule whose other atoms are significantly less electronegative than the atoms in the first molecule, will the oxidation number most likely increase (become less negative) or decrease (become more negative)?

3. Why must reduction always accompany oxidation?

4. You hook a wire to the anode and cathode of a battery. Which way do the electrons flow (cathode to anode or vice versa)?

5. Is an alkaline battery rechargeable? Why or why not?

6. A lead-acid battery uses lead, sulfuric acid, and what other reactant?

7. Give the oxidation numbers of all atoms in the following compounds:

a. MnO_2 b. PO_4^{3-} c. SO_2 d. $MgBr_2$ e. $KClO_3$ f. PF_4 g. $PtCl_4^{2-}$ h. N_2

8. An atom changes its oxidation number from -1 to +1. Is it oxidized or reduced? How many electrons did it take to do this?

9. An atom changes its oxidation number from 0 to -2. Is it oxidized or reduced? How many electrons did it take to do this?

10. An atom changes its oxidation number from +3 to -1. Is it oxidized or reduced? How many electrons did it take to do this?

11. An atom changes its oxidation number from 0 to +3. Is it oxidized or reduced? How many electrons did it take to do this?

12. Determine whether or not each of the following is a redox reaction. If it is, determine which atom is being reduced and which is being oxidized.

a. $O_2\ (g) + Cl_2\ (g) \rightarrow 2OCl\ (g)$
b. $2H_3O^+\ (aq) + Cu\ (s) + 2HNO_3\ (aq) \rightarrow Cu^{2+}\ (aq) + 2NO_2\ (g) + 4H_2O\ (l)$
c. $HNO_3\ (aq) + NH_3\ (aq) \rightarrow NH_4^+\ (aq) + NO_3^-\ (aq)$
d. $5HSO_3^-\ (aq) + 2IO_3^-\ (aq) \rightarrow I_2\ (aq) + 5SO_4^{2-}\ (aq) + 3H^+\ (aq) + H_2O\ (l)$
e. $Al_2O_3\ (aq) + 6NaOH\ (aq) \rightarrow 2Al(OH)_3\ (s) + 3Na_2O\ (aq)$

(There are two more problems on the next page.)

13. A Galvanic cell runs on the following reaction:

$$3Ba\ (s)\ +\ 2Al^{3+}\ (aq)\ \rightarrow\ 3Ba^{2+}\ (aq)\ +\ 2Al\ (s)$$

Draw a diagram for this Galvanic cell, labeling the electron flow, the anode and cathode, and the positive and negative sides of the Galvanic cell.

14. A Galvanic cell runs on the following reaction:

$$8H_2S\ (aq)\ +\ 8F_2\ (aq) \rightarrow\ 16H^+\ +\ 16F^-\ (aq) + S_8\ (s)$$

Draw a diagram for this Galvanic cell, labeling the electron flow, the anode and cathode, and the positive and negative sides of the Galvanic cell.

SOLUTIONS TO THE MODULE #1 TEST

1. (3 pts) The prefix "milli" means 0.001. The prefix "centi" means 0.01. The prefix "kilo" means 1,000.

2. (1 pt) To properly compare these measurements, we need to get them into the same units. I will convert mg to g, although you could just as easily convert g to mg.

$$\frac{30.0\ \cancel{mg}}{1} \times \frac{0.001\ g}{1\ \cancel{mg}} = 0.0300\ g$$

Since 30.0 mg is really equal to 0.0300 g, it is the smallest of the two masses. Thus, the 0.3 gram rock is the heaviest.

3. (1 pt, ½ for each) Remember, precision is determined by how many decimal places there are. The more decimal places, the more precise the number. Accuracy, on the other hand, tells us how close to the true value a measurement is. Since a football field is supposed to be 100.0 yards long, (c) is the most precise measurement, but (b) is the most accurate.

4. (1 pt) If the bobber floats on the surface of the water, its density must be less than that of water.

5. (1 pt) It takes 5 dashes to go from 10 to 20 mL. This must mean that each dash is worth 2 mL. Since the meniscus is between the first and second dash above the 20 mL mark, the answer is somewhere between 22 and 24 mL. The meniscus looks just slightly below halfway between the two, so we could estimate that it is a little less than 23 mL. Since we are always to estimate one more decimal place than the scale reads, we could say that the volume is 22.8 mL (any number between 22.6 and 23.2 would be fine)

6. (1 pt) Density is mass divided by volume. The volume is provided in the answer to question #5, so we take 32.13 g and divide by the volume given in #5 (22.8 mL), we get 1.41 grams per mL. (If the answer in #5 is different, this answer will be different. Check it by taking the mass and dividing by the answer the student gave for #5. There should only be three significant figures in any answer to this question.)

7. (1 pt) $\frac{16.2\ \cancel{m}}{1} \times \frac{1\ cm}{0.01\ \cancel{m}} = 1.62 \times 10^3\ cm$ The answer can also be given as 1,620 cm.

8. (1 pt) $\frac{345.6\ \cancel{mg}}{1} \times \frac{0.001\ g}{1\ \cancel{mg}} = 0.3456\ g$

9. (2 pts, one for each step) To do this problem, you cannot directly convert yards to miles, because you have no direct relationship between the two. Thus, you must first convert feet. Then you can convert to miles.

$$\frac{100.0\ \cancel{yards}}{1} \times \frac{3\ feet}{1\ \cancel{yard}} = 300.0\ feet$$

We need to keep all of the significant figures that were in 100.0, since the relationship between yards and feet is exact. Now we can convert to miles:

$$\frac{300.0 \text{ \sout{feet}}}{1} \times \frac{1 \text{ mile}}{5.280 \times 10^3 \text{ \sout{feet}}} = \underline{0.05682 \text{ miles}}$$

10. (1 pt) To test whether or not this nugget is gold, we simply compute the density. If the density is 19.3 g/mL, the nugget is gold. If not, the nugget is not gold.

$$\rho = \frac{m}{V} = \frac{661 \text{ g}}{34.2 \text{ mL}} = 19.3 \frac{\text{g}}{\text{mL}}$$

The nugget is gold!

11. (2 pts, one for converting to get consistent units, and one for the answer) The density is given in g/cc. Now remember, this is the same as g/mL, but we will go ahead and use cc's. Since density is mass divided by volume, if we know density and mass, we can get volume. Notice, however, that there are conflicting units. The density uses grams (g/cc), but the mass is given in kg. Thus, before I solve the problem, I need to convert from kg to g:

$$\frac{0.123 \text{ \sout{kg}}}{1} \times \frac{1000 \text{ g}}{1 \text{ \sout{kg}}} = 123 \text{ g}$$

Now that all units agree, we can use the equation for density to solve for volume:

$$\rho = \frac{m}{V}$$

$$11.4 \frac{\text{g}}{\text{cc}} = \frac{123 \text{ g}}{V}$$

$$V = \frac{123 \text{ g}}{11.4 \frac{\text{g}}{\text{cc}}} = \underline{10.8 \text{ cc}}$$

12. (1 pt) To get the decimal point to the right of the first digit, we need to move it 3 spaces to the left. Thus, the exponent on the ten will be a 3. Since this number is large, the exponent is positive. Thus, the answer is $\underline{3.478 \text{ x } 10^3}$.

13. (1 pt) There is an exponent of 4 on the ten. This means that we must move the decimal 4 places. Since the exponent is negative, we need to move the decimal so that the number is less than one. Thus, the answer is 0.0001245.

14. (3 pts, one for the volume in m^3, one for the conversion to cm^3, and one for the conversion to liters) This is the hardest problem on the test. First, we need to plug the radius in the equation and get the volume:

$$V = \frac{4}{3} \cdot (3.1416) \cdot (3.1 \text{ m})^3 = 1.2 \times 10^2 \text{ m}^3$$

There can only be two significant figures in the answer, because 3.1 has only two significant figures. That's why the number is rounded to 1.2 x 10^2 m^3, or 120 m^3. Remember, integers are exact, so we do not worry about the "4" or the "3" when counting significant figures.

Although this **is** the volume of the sphere, it is not the answer, because the question asked for the volume in *liters*. Do we know of a way to convert from m^3 to liters? No, because we don't know of a relationship between them. We do know, however, that a mL is the same as a cm^3. Thus, we can convert from m^3 to cm^3, which then is the same as mL.

$$\frac{1.2 \times 10^2 \text{ m}^3}{1} \times \left(\frac{1 \text{ cm}}{0.01 \text{ m}}\right)^3$$

$$\frac{1.2 \times 10^2 \cancel{\text{m}^3}}{1} \times \frac{1 \text{ cm}^3}{0.000001 \cancel{\text{m}^3}} = 1.2 \times 10^8 \text{ cm}^3 = 1.2 \times 10^8 \text{ mL}$$

Now that we have the volume in mL, we can convert to L:

$$\frac{1.2 \times 10^8 \cancel{\text{mL}}}{1} \times \frac{0.001 \text{ L}}{1 \cancel{\text{mL}}} = \underline{1.2 \times 10^5 \text{ L}}$$

15. (1 pt) The mass would be the same. Unlike weight, mass never changes, regardless of where you are.

Total possible points: 21

SOLUTIONS TO THE MODULE #2 TEST

1. (1 pt) In order to do work, you must create motion. If no motion is created, there is no work.

2. (2 pts, one for the first law, and one for the candle explanation) The First Law of Thermodynamics states that energy cannot be created or destroyed. It can only change forms. Thus, the heat in the flame of a burning candle was always there, it was simply stored in the candle and the surrounding oxygen. Only after the candle was lit could the energy be released.

3. (2 pts, one for the difference between the two, and one for the examples) Kinetic energy is energy in motion. Potential energy is energy that has been stored. Heat, lightning, and flames are all examples of kinetic energy. All matter has stored energy, so any substance would be an example of potential energy.

4. (1 pt) A food Calorie is equal to 1,000 chemistry calories.

5. (1 pt) If an object's temperature increases, it must be gaining energy.

6. (1 pt) Energy is measured in Joules and calories. 1 cal = 4.184 J.

7. (2 pts, one for the answer in °C, the other for the answer in °F)

$$^{\circ}C = K - 273.15 = 100.0 - 273.15 = \underline{-173.2\ ^{\circ}C}$$

Since we are adding, we look at decimal place. The 100.0 has its last significant figure in the tenths place, so the answer can only go out to the tenths place.

$$^{\circ}F = \frac{9}{5} \cdot (^{\circ}C) + 32$$

$$^{\circ}F = \frac{9}{5} \cdot (-173.2) + 32 = \underline{-279.8\ ^{\circ}F}$$

All numbers in this problem except the temperature in °C are exact. Thus, we keep the same number of significant figures as we had in the original measurement.

8. (1 pt) The lead has the smallest specific heat, so it is easiest to heat up. Therefore, the lead will be the hottest.

9. (1 pt) Water boils at 100 °C, and while it boils its temperature stays constant.

10. (2 pts, one for converting to make the units consistent, and one for the answer) This is an application of Equation (2.3). The only problem is that there is a unit discrepancy here. The mass is in kg, but the specific heat uses g. Thus, we need to convert one of those numbers. I will convert the mass from kg to g:

$$\frac{10.0\ \cancel{kg}}{1} \times \frac{1000\ g}{1\ \cancel{kg}} = 1.00 \times 10^{4}\ g$$

Now that the units agree, we can use Equation (2.3):

$$q = m \cdot c \cdot \Delta T = (1.00 \times 10^4\ g) \cdot (0.3851\ \frac{J}{g \cdot {}^\circ C}) \cdot (175.0\ {}^\circ C - 25.0\ {}^\circ C)$$

$$q = (1.00 \times 10^4\ \cancel{g}) \cdot (0.3851\ \frac{J}{\cancel{g} \cdot {}^\circ \cancel{C}}) \cdot (150.0\ {}^\circ \cancel{C}) = \underline{5.78 \times 10^5\ J}$$

Notice that there are only three significant figures because the mass has only three.

11. (2 pts, one for converting to make the units consistent, and one for the answer) This is an application of Equation (2.3). The only problem is that there is a unit discrepancy here. The mass is in kg, but the specific heat uses g. Thus, we need to convert one of those numbers. I will convert the mass from kg to g:

$$\frac{1.50\ \cancel{kg}}{1} \times \frac{1000\ g}{1\ \cancel{kg}} = 1.50 \times 10^3\ g$$

Now that the units agree, we can use Equation (2.3). Remember, though, that when an object releases heat, the heat is negative. Thus, we will use -5,505 cal as "q" in the equation:

$$q = m \cdot c \cdot \Delta T$$

$$-5{,}505\ J = (1.50 \times 10^3\ g) \cdot (0.4521\ \frac{J}{g \cdot {}^\circ C}) \cdot (T_{final} - 110.0\ {}^\circ C)$$

$$(T_{final} - 110.0\ {}^\circ C) = \frac{-5{,}505\ \cancel{J}}{(1.50 \times 10^3\ \cancel{g}) \cdot (0.4521\ \frac{\cancel{J}}{\cancel{g} \cdot {}^\circ C})} = -8.12\ {}^\circ C$$

$$T_{final} = \underline{101.9\ {}^\circ C}$$

Notice that the difference in temperature (-8.12 °C) has only three significant figures. However, to get T_{final}, you must *add* 110.0 °C to that number. At that point, you do not count significant figures. Instead, you look at decimal place. Since the least precise number (110.0) has its last significant figure in the tenths place, the answer must be reported to the tenths place.

12. (1 pt) This is an application of Equation (2.3). There are no unit conflicts here, so we can just use the equation:

$$q = m \cdot c \cdot \Delta T$$

$$1{,}145\ J = (50.0\,g) \cdot (c) \cdot (10.0\ {}^\circ C)$$

$$c = \frac{1{,}145\ \text{J}}{(50.0\ \text{g}) \cdot (10.0\ ^{\circ}\text{C})} = 2.29\ \frac{\text{J}}{\text{g} \cdot {}^{\circ}\text{C}}$$

13. (1 pt) A calorimeter consists of an insulated container, a thermometer, a liquid (usually water), and an object that releases or absorbs heat. When the object releases heat into the water and calorimeter or takes heat away from them, the temperature changes. This change in temperature can be related to the amount of energy transferred with Equation (2.3).

14. (2 pts, one for q_{water}, and one for the answer) We can ignore the calorimeter here, so all we need to do is determine q_{water}:

$$q_{water} = m \cdot c \cdot \Delta T = (125\ \text{g}) \cdot (4.184\ \frac{\text{J}}{\text{g} \cdot {}^{\circ}\text{C}}) \cdot (24.5\ ^{\circ}\text{C} - 22.3\ ^{\circ}\text{C}) = (125\ \text{g}) \cdot (4.184\ \frac{\text{J}}{\text{g} \cdot \cancel{^{\circ}\text{C}}}) \cdot (2.2\ \cancel{^{\circ}\text{C}})$$

$$q_{water} = 1{,}200\ \text{J}$$

Now we can determine q_{metal}:

$$-q_{object} = q_{water} + q_{calorimeter}$$

$$q_{metal} = -1{,}200\ \text{J}$$

Now we can use Equation (2.3) to determine the specific heat.

$$q = m \cdot c \cdot \Delta T$$

$$-1{,}200\ \text{J} = (50.0\ \text{g}) \cdot (c) \cdot (24.5\ ^{\circ}\text{C} - 123.0\ ^{\circ}\text{C})$$

$$c = \frac{-1{,}200\ \text{J}}{(50.0\ \text{g}) \cdot (-98.5\ ^{\circ}\text{C})} = 0.24\ \frac{\text{J}}{\text{g} \cdot {}^{\circ}\text{C}}$$

15. (3 pts, one for q_{water}, one for $q_{calorimeter}$, and one for the answer) To figure out the mass of the glass, we need to figure out how much energy it released. We cannot ignore the calorimeter in this problem, so we must determine both q_{water} and $q_{calorimeter}$:

$$q_{water} = m \cdot c \cdot \Delta T = (75.0\ \text{g}) \cdot (4.184\ \frac{\text{J}}{\text{g} \cdot {}^{\circ}\text{C}}) \cdot (26.6\ ^{\circ}\text{C} - 24.3\ ^{\circ}\text{C}) = (75.0\ \text{g}) \cdot (4.184\ \frac{\text{J}}{\text{g} \cdot \cancel{^{\circ}\text{C}}}) \cdot (2.3\ \cancel{^{\circ}\text{C}})$$

$$q_{water} = 720\ \text{J}$$

$$q_{calorimeter} = m \cdot c \cdot \Delta T = (7.0\ g) \cdot (1.23\ \frac{J}{g \cdot {}^{o}C}) \cdot (26.6\ {}^{o}C - 24.3\ {}^{o}C) = (7.0\ \cancel{g}) \cdot (1.23\ \frac{J}{\cancel{g} \cdot {}^{o}\cancel{C}}) \cdot (2.3\ {}^{o}\cancel{C})$$

$$q_{calorimeter} = 2.0 \times 10^{1}\ J$$

Notice that the only way I could properly report $q_{calorimeter}$ was to use scientific notation, because it must have two significant figures. Now we can use the calorimetery equation:

$$-q_{object} = q_{water} + q_{calorimeter} = 720\ J + 2.0 \times 10^{1}\ J$$

$$q_{glass} = -740\ J$$

Now that we know the heat released by the glass, we can determine its mass:

$$q = m \cdot c \cdot \Delta T$$

$$-740\ J = (m) \cdot (0.8372\ \frac{J}{g \cdot {}^{o}C}) \cdot (26.6\ {}^{o}C - 90.0\ {}^{o}C)$$

$$m = \frac{-740\ \cancel{J}}{(0.8372\ \frac{\cancel{J}}{g \cdot {}^{o}\cancel{C}}) \cdot (-63.4\ {}^{o}\cancel{C})} = \underline{14\ g}$$

Total possible points: 23

SOLUTIONS TO THE MODULE #3 TEST

1. (1 pt) Elements cannot be decomposed, so it must be a compound.

2. (1 pt) Substances that conduct electricity when dissolved in water are ionic compounds.

3. (1 pt) A compound is always made up with the same proportion of its elements.

4. (1 pt) Compounds "a" and "b" are really the same compound. In order to make two different compounds, the elements would have to combine in different *proportions*. Since 48.0 g of oxygen + 42.0 g of nitrogen is just 3 times 16.0 g of oxygen and 14.0 g of nitrogen, the elements have combined in the same proportion, so they make the same compound.

5. (1 pt) It is a nonmetal.

6. (1 pt) Scientists are convinced that the discontinuous theory of matter is correct.

7. (1 pt) The existence of atoms has not been proven. Instead, we can understand so many of the facts we know by assuming that atoms exist that we take their existence on faith.

8. (1 pt) The law of mass conservation says that the total mass must always stay constant. Thus, if there was 100.1 g to start with, there must be 100.1 g after the decomposition. The carbon and magnesium add up to 52.1 g. Therefore, there must have been 48.0 g of oxygen.

9. (2 pts, one for each mass) Since there were 2.0 g of oxygen left over, the chemist only needed 48.0 g of oxygen to make 80.0 g of sulfur trioxide. Since there was no left over sulfur, the chemist needed all 32.0 g of sulfur. Thus, to make 80.0 g of sulfur trioxide, you need 32.0 g of sulfur and 48.0 g of oxygen. To make 100.0 g of sulfur trioxide, you simply need to scale up by $100.0 \div 80.0 = 1.25$. Thus, you need 1.25×32.0 g = 40.0 g of sulfur and 1.25×48.0 g = 60.0 g of oxygen.

10. (2 pts, one for each mass) Based on the recipe given, we have 10 times as much carbon and 5 times as much oxygen. Thus, we can only make 5 times as much carbon dioxide as in the recipe, because we will run out of oxygen long before carbon. This means we will make $5 \times 44.0 = 2.20 \times 10^2$ g of carbon dioxide. This will use up all of the oxygen, but a lot of carbon will be left over. Since we can only make 5 times the amount of carbon dioxide as is in the recipe, we will use only $5 \times 12.0 =$ 60.0 g of carbon. Since we started with 120.0 g of carbon and only used 60.0, there are 60.0 g of leftover carbon.

11. (1 pt) According to the formula, the molecule has 5 carbon atoms and 10 hydrogen atoms. Thus, there are already 15 atoms. With 17 total atoms, there must be 2 more. Thus $x = 2$.

12. (3 pts, one for each answer. Subtract one for every wrong element listed.) Metals are on the left of the jagged line of the periodic chart, nonmetals are on the right. Thus Ne, Xe, and S are nonmetals.

13. (2 pts, one for each answer. Subtract one for every wrong element listed.) Ionic compounds conduct electricity when dissolved in water. Ionic compounds have at least one metal and one nonmetal in them. Thus, $NaNO_3$ and $BaSO_4$ will conduct electricity when dissolved in water.

14. Ionic compounds do not use prefixes. Covalent compounds do. Thus (a) and (d) will not have prefixes, (b) and (c) will.

a. (1 pt) aluminum sulfide
b. (1 pt) sulfur difluoride
c. (1 pt) dicarbon hexafluoride
d. (1 pt) beryllium fluoride

15. (1 pt) SO_3

Total possible points: 23

SOLUTIONS TO THE MODULE #4 TEST

1. (1 pt, ¼ point for each) Element, Compound, Heterogeneous Mixture, Homogeneous Mixture

2. (1 pt, ½ point for each) Physical, Chemical

3. (1 pt) A pure substance.

4. a. (1 pt) heterogeneous mixture
   b. (1 pt) element
   c. (1 pt) compound
   d. (1 pt) homogeneous mixture

5. (3 pts, one for each answer) Solid HCl occupies less space, its molecules are closer together, and they move slower.

6. (1 pt) It was cooled.

7. (1 pt) Water is one of the few substances in which its solid phase is less dense than its liquid. That's why ice floats on water. For the vast majority of other compounds, the solid phase is more dense than the liquid phase, so frozen HNO_3 does not float in liquid HNO_3.

8. a. (1 pt) Just like cutting your hair, this change can be reversed by gluing each blade of grass back together. Thus, this is a physical change.

   b. (1 pt) The turkey can be re-frozen, so this is a physical change.

   c. (1 pt) You can't reverse burning. This is a chemical change.

9. (2 pts, one for each answer. Subtract a point for every wrong element listed) The two homonuclear diatomic elements on this list are nitrogen: N_2 and fluorine: F_2

10. (1 pt) Yes

11. (1 pt) $2H_2O\ (l) + 2Na\ (s) \rightarrow 2NaOH\ (aq) + H_2\ (g)$

12. (1 pt) $2Cl_2\ (g) + C_2H_2\ (g) \rightarrow C_2H_2Cl_4\ (l)$

13. (1 pt) $C_9H_{20} + 14O_2 \rightarrow 9CO_2 + 10H_2O$

14. (1 pt) $C_2H_6O\ (l) + 3O_2\ (g) \rightarrow 2CO_2\ (g) + 3H_2O\ (g)$

15. (1 pt) If a chemical equation is balanced, then the same number of atoms appear on both sides. This would mean the mass is the same whether we are talking about reactants or products. That's another way of stating the Law of Mass Conservation.

Total possible points: 23

SOLUTIONS TO THE MODULE #5 TEST

1. (1 pt) Decomposition, because a single compound is breaking down into its elements.

2. (1 pt) None of these

3. (1 pt) Combustion, because oxygen is being added and CO_2 and H_2O are being produced.

4. (1 pt) Formation, because elements are forming a single compound.

5. (1 pt) Remember, hydrogen and bromine are both homonuclear diatomics, so their formulas are H_2 and Br_2: $\underline{2P + 2H_2 + Br_2 \rightarrow 2PH_2Br}$

6. (1 pt) Remember, oxygen is a homonuclear diatomic: $\underline{Na_2CrO_4 \rightarrow 2Na + Cr + 2O_2}$

7. (2 pts, one for writing an equation that has all of the right reactants and products, and one for balancing the equation) $\underline{2C_{10}H_{22}\ (l) + 31O_2\ (g) \rightarrow 20CO_2\ (g) + 22H_2O\ (g)}$

8. (2 pts, one for each conversion) According to the periodic chart, Pr has a mass of 140.9 amu. We just need to convert it to kg. Unfortunately, the relationship we were given is between amu and grams. Thus, we need to do a two-step conversion.

$$\frac{140.9\ \cancel{\text{amu}}}{1} \times \frac{1.66 \times 10^{-24}\ \cancel{\text{g}}}{1.00\ \cancel{\text{amu}}} \times \frac{1\ \text{kg}}{1{,}000\ \cancel{\text{g}}} = \underline{2.34 \times 10^{-25}\ \text{kg}}$$

9. (1 pt) To get the mass of a molecule, we just add the masses of all atoms in the molecule:

$$1.01\ \text{amu} + 35.5\ \text{amu} + 4\text{x}16.0\ \text{amu} = \underline{100.5\ \text{amu}}$$

Since we are adding here, we look at decimal place. The two least precise numbers (35.5 and 16.0) each have their last significant figure in the tenths place, so the answer must be reported to the tenths place.

10. (2 pts, one for getting the mass, and one for the answer) To get a relationship between grams and moles, we get the mass of the molecule:

$$23.0\ \text{amu} + 79.9\ \text{amu} + 3\text{x}16.0\ \text{amu} = 150.9\ \text{amu}$$

This means 1 mole = 150.9 grams. Now we can use this in a conversion:

$$\frac{34.5\ \cancel{\text{g}}}{1} \times \frac{1\ \text{mole}}{150.9\ \cancel{\text{g}}} = \underline{0.229\ \text{moles}}$$

11. (2 pts, one for getting the mass, and one for the answer) To get a relationship between grams and moles, we get the mass in amu:

$$2\text{x}14.0\ \text{amu} + 4\text{x}1.01\ \text{amu} + 3\text{x}16.0\ \text{amu} = 80.0\ \text{amu}$$

This means 1 mole = 80.0 g. Now we can convert from moles to grams:

$$\frac{12.3\ \cancel{\text{moles}}}{1} \times \frac{80.0\text{ g}}{1\ \cancel{\text{mole}}} = \underline{984\text{ g}}$$

12. (2 pts, one for the equation, one for the answer) To determine this, we need to figure out the decomposition reaction:

$$2K_2CO_3 \rightarrow 4K + 2C + 3O_2$$

Thus, 2 moles of K_2CO_3 = 3 moles of O_2.

$$\frac{0.45\ \cancel{\text{moles } K_2CO_3}}{1} \times \frac{3\text{ moles } O_2}{2\ \cancel{\text{moles } K_2CO_3}} = \underline{0.68\text{ moles } O_2}$$

Thus, <u>0.68 moles O_2</u> are formed.

13. (2 pts, one for converting to moles, and one for the answer) To count atoms, we need to know how many moles. Thus, we need to take both of these masses and convert them to moles:

$$\frac{150.0\ \cancel{\text{g Au}}}{1} \times \frac{1\text{ mole Au}}{197.0\ \cancel{\text{g Au}}} = 0.7614\text{ moles Au}$$

$$\frac{10.0\ \cancel{\text{g Li}}}{1} \times \frac{1\text{ mole Li}}{6.94\ \cancel{\text{g Li}}} = 1.44\text{ moles Li}$$

Thus, <u>10.0 g of lithium</u> has more atoms than 150.0 g of Au, because it has more moles.

14. (1 pt) <u>You should be worried about incomplete combustion producing carbon monoxide</u>.

15. (1 pt) Since we have one mole, the mass of the sample in grams is also the mass of the molecule in amu. Thus, we know that this molecules mass is 111.1 amu. Since it has 2 chlorine atoms, the chlorines alone take up 71.0 amu of that mass. Thus, 40.1 amu are left over for X. If we look at the periodic chart, the only atom with a mass of 40.1 amu is Ca. Thus <u>X = Ca</u>.

Total possible points: 21

SOLUTIONS TO THE MODULE #6 TEST

1. (1 pt) A stoichiometric coefficient is the number to the left of a substance in a chemical equation.

2. (1 pt) 1,2,1

3. (1 pt) Nothing. The amount of limiting reactant determines the amount of product formed.

4. (1 pt) Gay-Lussac's Law states that the stoichiometric coefficients can be used to relate the volumes of gases in a chemical equation. The limitation is that it only works for gases.

5. (3 pts, one for writing a decomposition equation, one for converting to moles, and one for the answer) The unbalanced equation for this decomposition reaction is:

$$Na_xP_yO_z \rightarrow Na + P + O_2$$

We can now determine how many moles of each product was made:

$$\frac{6.9\ \cancel{g\ Na}}{1} \times \frac{1\ \text{mole Na}}{23.0\ \cancel{g\ Na}} = 0.30\ \text{moles Na}$$

$$\frac{3.1\ \cancel{g\ P}}{1} \times \frac{1\ \text{mole P}}{31.0\ \cancel{g\ P}} = 0.10\ \text{moles P}$$

$$\frac{6.4\ \cancel{g\ O_2}}{1} \times \frac{1\ \text{mole } O_2}{32.0\ \cancel{g\ O_2}} = 0.20\ \text{moles } O_2$$

Thus,

$$Na_xP_yO_z \rightarrow 0.30\ Na + 0.10\ P + 0.20\ O_2$$

Dividing by the smallest to make integers:

$$Na_xP_yO_z \rightarrow 3Na + P + 2O_2$$

In order to get the equation to balance, then, x = 3, y = 1, and z = 4. The empirical formula, then, is Na_3PO_4.

6. (1 pt) The chemical equation relates moles of one substance to moles of another.

$$\frac{2.3\ \cancel{\text{moles NaOH}}}{1} \times \frac{2\ \text{moles Al(OH)}_3}{6\ \cancel{\text{moles NaOH}}} = 0.77\ \text{moles Al(OH)}_3$$

7. (3 pts, one for each of the equations below)

$$\frac{15.0\ \cancel{g\ AgNO_3}}{1} \times \frac{1\ \text{mole}\ AgNO_3}{169.9\ \cancel{g\ AgNO_3}} = 8.83 \times 10^{-2}\ \text{moles}\ AgNO_3$$

$$\frac{8.83 \times 10^{-2}\ \cancel{\text{moles}\ AgNO_3}}{1} \times \frac{1\ \text{mole}\ CaBr_2}{2\ \cancel{\text{moles}\ AgNO_3}} = 4.42 \times 10^{-2}\ \text{moles}\ CaBr_2$$

$$\frac{4.42 \times 10^{-2}\ \cancel{\text{moles}\ CaBr_2}}{1} \times \frac{199.9\ g\ CaBr_2}{1\ \cancel{\text{mole}\ CaBr_2}} = \underline{8.84\ g\ CaBr_2}$$

8. (2 pts, one for determining that it is not empirical, one for the empirical formula) An empirical formula has the lowest possible numbers as subscripts. The subscripts cannot be divisible by a common factor. These subscripts are both divisible by 2. Thus, it is not an empirical formula. To get the empirical formula, you need to divide by the largest common divisor. Therefore, $AlCl_3$ is the empirical formula.

9. (3 pts, one for each of the equations below)

$$\frac{50.0\ \cancel{g\ HClO_3}}{1} \times \frac{1\ \text{mole}\ HClO_3}{84.5\ \cancel{g\ HClO_3}} = 0.592\ \text{moles}\ HClO_3$$

$$\frac{0.592\ \cancel{\text{moles}\ HClO_3}}{1} \times \frac{6\ \text{mole}\ ClO_2}{5\ \cancel{\text{moles}\ HClO_3}} = 0.710\ \text{moles}\ ClO_2$$

$$\frac{0.710\ \cancel{\text{moles}\ ClO_2}}{1} \times \frac{67.5\ g\ ClO_2}{1\ \cancel{\text{mole}\ ClO_2}} = \underline{47.9\ g\ ClO_2}$$

10. (1 pt) Since everything that we are interested in is now a gas, we can use Gay-Lussac's Law,

$$\frac{10.0\ \cancel{L\ ClO_2}}{1} \times \frac{1\ L\ HCl}{6\ \cancel{L\ ClO_2}} = \underline{1.67\ L\ HCl}$$

11. (1 pt) You cannot solve it because $HClO_3$ is a liquid, not a gas. Since you are only given the amount of ClO_2 in terms of volume, then the only way to solve it is by using Gay-Lussac's law. Unfortunately, this only works when all substances of interest are gasses.

12. (1 pt) The empirical formula tells us that for every one C there must also be two H's. To keep this relationship but still get the mass to add to 70.1 amu, the formula must be C_5H_{10}.

13. (1 pt) The word "gram" should be replaced by "mole" (or "molecule") each time it appears. Stoichiometric coefficients only relate moles (or molecules), not grams.

14. (1 pt) According to the equation, 3 moles of CO react with 1 mole of Fe_2O_3. Thus, when the 3 moles of CO are exhausted, there will still be an extra mole of Fe_2O_3. This means that CO is the limiting reactant

15. (1 pt) Since CO is the limiting reactant, we use it in our calculation. According to the equation, 3 moles of CO make 2 moles Fe.

Total possible points: 22

SOLUTIONS TO THE MODULE #7 TEST

1. (1 pt) They repel each other.

2. (1 pt) Since there are as many positive charges as negative ones, it is electrically neutral.

3. (1 pt) Crookes discovered cathode rays, which were later determined to be electrons.

4. (2 pts, one for the number of protons and electrons, one for the number of neutrons) The periodic chart tells us that Ba has an atomic number of 56. This means there are 56 protons. Since Ba has no charge, it must also have 56 electrons. The mass number (141) is the sum of protons plus neutrons. If there are 56 protons, then there must be 85 neutrons to make the sum 141. Thus, there are 56 protons, 56 electrons, and 85 neutrons.

5. (2 pts, one for the abbreviation, and one for the mass number) The abbreviation is determined by the number of protons. Since there are 45 protons, we are looking for element 45 on the periodic chart, which is Rh. Since there are also 45 electrons, there is no charge. The mass number is the sum of protons plus neutrons, or 102. Thus, the atom is ^{102}Rh.

6. (1 pt) Isotopes are atoms with the same number of protons but different numbers of neutrons.

7. (2 pts, one for amplitude, and one for frequency) Brightness is determined by amplitude; therefore, they have the same amplitude. According to our mnemonic ROY G. BIV, the indigo light has a shorter wavelength than the yellow light. Shorter wavelength means higher frequency. The indigo one emits light with a higher frequency.

8. (2 pts, one for making the units consistent, and one for the answer) This is an application of Equation (7.1), but the units disagree. The speed of light was given using meters (m/s), but the wavelength is in nm. Thus, we must first convert nm to m:

$$\frac{10.0\ \cancel{\text{nm}}}{1} \times \frac{1 \times 10^{-9}\ \text{m}}{1\ \cancel{\text{nm}}} = 1.00 \times 10^{-8}\ \text{m}$$

Now that the units agree, we can use Equation (7.1):

$$f = \frac{c}{\lambda} = \frac{3.0 \times 10^{8}\ \frac{\cancel{\text{m}}}{\text{s}}}{1.00 \times 10^{-8}\ \cancel{\text{m}}} = 3.0 \times 10^{16}\ \frac{1}{\text{s}}$$

Since 1/s is the same as Hz, the frequency is 3.0×10^{16} Hz.

9. (2 pts, one for frequency, and one for wavelength) There is no direct relationship between energy and wavelength. However, given the energy, we can get the frequency (Equation (7.2)), and then we can use Equation (7.1) to go from frequency to wavelength.

$$E = h \cdot f$$

$$2.3\times10^{-15}\ \text{J} = (6.63\times10^{-34}\ \frac{\text{J}}{\text{Hz}})\cdot \text{f}$$

$$\text{f} = \frac{2.3\times10^{-15}\ \cancel{\text{J}}}{6.63\times10^{-34}\ \frac{\cancel{\text{J}}}{\text{Hz}}} = 3.5\times10^{18}\ \text{Hz}$$

Now that we have frequency, we can get wavelength.

$$\text{f} = \frac{\text{c}}{\lambda}$$

$$3.5\times10^{18}\ \text{Hz} = \frac{3.0\times10^{8}\ \frac{\text{m}}{\text{s}}}{\lambda}$$

$$\lambda = \frac{3.0\times10^{8}\ \frac{\text{m}}{\cancel{\text{s}}}}{3.5\times10^{18}\ \frac{1}{\cancel{\text{s}}}} = 8.6\times10^{-11}\ \text{m}$$

The wavelength is $8.6\text{x}10^{-11}$ m.

10. (1 pt) The neutron is the heaviest of the three.

11. (1 pt) Since the atom is emitting light, its electrons are losing energy. Its electrons are jumping from high energy orbitals to low energy ones.

12. (1 pt) The s orbitals have the lowest energy, the p orbitals have the next highest energy, and the d orbitals have the highest energy.

13. (1 pt) $2s^3$ is impossible. There can only be 2 electrons in an s orbital.

14. a. (1 pt) $1s^22s^22p^63s^23p^64s^2$

b. (1 pt) $1s^22s^22p^63s^23p^64s^23d^{10}4p^65s^24d^3$

c. (1 pt) $1s^22s^22p^63s^23p^64s^23d^6$

15. a. (1 pt) $[Rn]7s^1$ b. (1 pt) $[Ne]3s^23p^5$ c. (1 pt) $[Kr]5s^24d^{10}$

Total possible points: 24

SOLUTIONS TO THE MODULE #8 TEST

1. (3 pts, one for each element. Subtract one point for each wrong element listed.) P, N, and As, since they are in the same column of the periodic chart.

2. (1 pt) Valence electrons exist in the orbital(s) at the highest energy level. The orbital(s) at the highest energy level have the largest number in front of them. In this case, 4s and 4p have the largest number in front, so there are four valence electrons.

3. (1 pt) The "B" columns.

4. (1 pt) 8A

5. (1 pt) Metal ions always develop a positive charge, because they give away electrons.

6. (1 pt) covalent

7. (1 pt) Ionization potential

8. (1 pt) Ozone blocks ultraviolet rays that come from the sun.

9. (1 pt)

```
•Al•
 •
```

10. a. (1 pt) In ionic compounds K develops a +1 charge because it is in group 1A. As develops a -3 charge because it is in group 5A. Switching the charges and ignoring the signs gives us: K_3As

b. (1 pt) The Roman numeral tells us that Fe develops a +3 charge. O develops a -2 charge because it is in group 6A. Switching the charges and ignoring the signs gives us: Fe_2O_3

c. (1 pt) In ionic compounds Ca develops a +2 charge because it is in group 2A. S develops a -2 charge because it is in group 6A. Since the charges are the same we ignore them: CaS

11. (1 pt) Na, because it is farthest to the left on the periodic chart.

12. (2 pts) Ba < Sr < Ca < Mg, because as you go down the chart, electronegativity decreases.

13. a. (1 pt)

```
   H
   |
 H-P:
   |
   H
```

b. (1 pt)

```
 ••   ••
:Br - Br:
 ••   ••
```

c. (1 pt) SiH_4

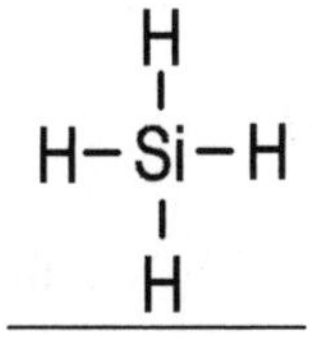

d. (1 pt) $Cl-O-Cl$

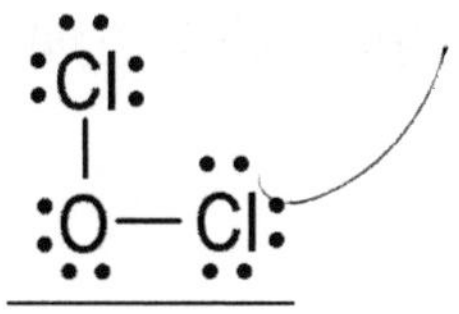

e. (1 pt) $O=Si=O$

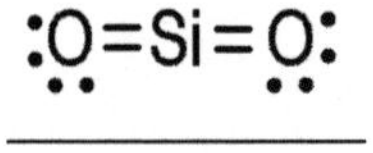

f. (1 pt) $S=C=S$

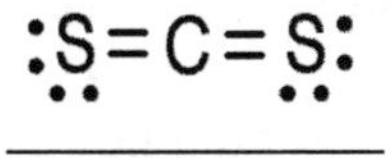

g. (1 pt) $F-Si\equiv N$

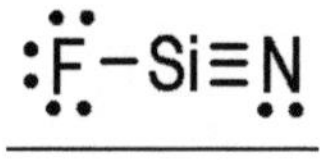

Another acceptable Lewis structure is:

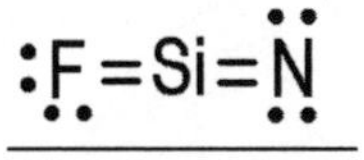

$F=Si=N$

Total possible points: 24

SOLUTIONS TO THE MODULE #9 TEST

1. (1 pt) The CO_3 has three oxygens, but it is in a set of parentheses with a 3 as a subscript. This means that there are three CO_3's. Thus, there are 3x3 = 9 oxygens.

2. (1 pt) Since polyatomic ions have a groups of atoms that share electrons, they are like covalent compounds. However, since they also contain extra electrons or a deficit of electrons, they are also like ionic compounds.

3. a. (1 pt) $Ca(C_2H_3O_2)_2$ b. (1 pt) $BaCrO_4$ c. (1 pt) $(NH_4)_2CO_3$

4. (1 pt) $2NH_4Cl\ (aq) + Ba(NO_3)_2\ (aq) \rightarrow BaCl_2\ (s) + 2NH_4NO_3\ (aq)$

5. (1 pt) Non-bonding electron pairs repel more strongly than bonds.

6. (1 pt) Molecules do anything they can to get their central atom's valence electron groups far away from each other. They can often get them farther apart if they use all three dimensions of space.

7. (2 pts, one for the shape and bond angle, one for the picture) Linear, with a bond angle of 180°.

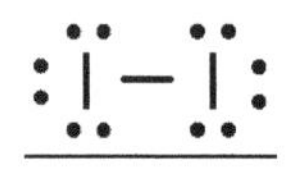

8. (2 pts, one for the shape and bond angle, one for the picture) Pyramidal, with a bond angle of 107°.

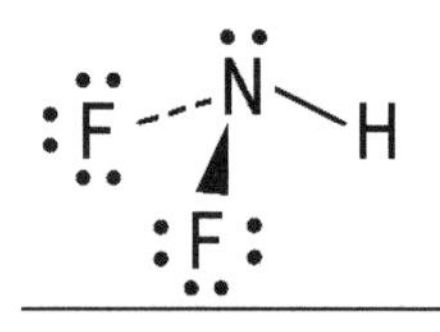

9. (2 pts, one for the shape and bond angle, one for the picture) Tetrahedral, with a bond angle of 109°.

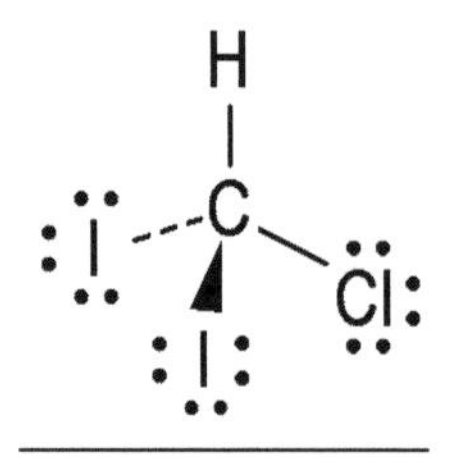

10. (2 pts, one for the shape and bond angle, one for the picture) Bent, with a bond angle of 105^0.

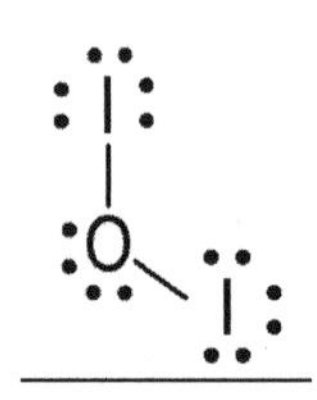

11. (2 pts, one for the shape and bond angle, one for the picture) Linear, with a bond angle of 180°.

:S̤=Si=S̤:

12. a. (1 pt) polar covalent
    b. (1 pt) polar covalent
    c. (1 pt) polar covalent
    d. (1 pt) purely covalent
    e. (1 pt) ionic
    f. (1 pt) purely covalent
    g. (1 pt) purely covalent

13. (4 pts, one for each letter. Subtract 1 point for every incorrect letter listed.) Since water is polar, only ionic or other polar compounds will dissolve in it. Thus, (a), (b), (c), and (e) should dissolve in water.

Total possible points: 29

SOLUTIONS TO THE MODULE #10 TEST

1. (1 pt) An acid

2. Acids donate an H^+, whereas bases receive it.

   a. (1 pt) Acid - C_2H_6O Base - CO_3^{2-}

   b. (1 pt) Acid - HNO_3 Base - H_2O

   c. (1 pt) Acid - HCl Base - KOH

3. (1 pt) A base is any substance that accepts H^+ ions.

4. (1 pt, subtract 1 point if an incorrect letter is listed) Diprotic acids have 2 H^+'s to donate. Thus, b is a diprotic acid.

5. (1 pt) Amphiprotic substances can act as either acids or bases. Water is a common example.

6. (1 pt) To determine whether it is an indicator or not, mix it with several acids and bases. If it is always the same color in acids and always a different color in bases, it is an indicator.

7. (1 pt) The endpoint of a titration indicates that the number of moles of acid added are enough to eat up any base present. If the chemical equation indicates a 2:1 mole ratio, then the number of moles are not equal, they are a factor of two apart.

8. (2 pts, one for the correct substances, one for the equation being balanced)
$H_3PO_4 + 3CsOH \rightarrow Cs_3PO_4 + 3H_2O$

9. (2 pts, one for the correct substances, one for the equation being balanced)
$2HCl + Be(OH)_2 \rightarrow BeCl_2 + 2H_2O$

10. (1 pt) $HF + H_2O \rightarrow H_3O^{1+} + F^{1-}$

11. (2 pts, one for both conversions, and one for the answer) Concentration is number of moles divided by number of liters. Thus, I must first convert from grams to moles:

$$\frac{12.23\ \cancel{g\ NH_3}}{1} \times \frac{1\text{ mole }NH_3}{17.0\ \cancel{g\ NH_3}} = 0.719\text{ moles }NH_3$$

Now I can divide by the volume, making sure I convert 560.0 mL into 0.5600 L first:

$$\text{concentration} = \frac{\#\text{ moles}}{\#\text{ Liters}} = \frac{0.719\text{ moles}}{0.5600\text{ L}} = \underline{1.28\text{ M}}$$

12. (1 pt) This is a dilution problem, so we use the dilution equation.

$$M_1 \cdot V_1 = M_2 \cdot V_2$$

$$(10.0 \text{ M}) \cdot V_1 = (3.0 \text{ M}) \cdot (450.0 \text{ mL})$$

$$V_1 = \frac{(3.0 \cancel{\text{M}}) \cdot (450.0 \text{ mL})}{(10.0 \cancel{\text{M}})} = 1.4 \times 10^2 \text{ mL}$$

<u>The chemist must take 1.4×10^2 mL of the stock solution and add it to enough water to make 450.0 mL of solution</u>.

13. (3 pts, one for the balanced equation, and two for the stoichiometry) To solve this stoichiometry problem, we first must get the chemical equation:

$$2HCl + Mg(OH)_2 \rightarrow 2H_2O + MgCl_2$$

Now we can do stoichiometry:

$$\text{moles HCl} = 5.6 \frac{\text{moles}}{\cancel{\text{L}}} \times 0.0300 \cancel{\text{L}} = 0.17 \text{ moles HCl}$$

$$\frac{0.17 \cancel{\text{moles HCl}}}{1} \times \frac{1 \text{ mole MgCl}_2}{2 \cancel{\text{moles HCl}}} = 0.085 \text{ moles MgCl}_2$$

$$\frac{0.085 \cancel{\text{moles MgCl}_2}}{1} \times \frac{95.3 \text{ g MgCl}_2}{1 \cancel{\text{mole MgCl}_2}} = \underline{8.1 \text{ g MgCl}_2}$$

14. (3 pts, one for the balanced equation, and two for the stoichiometry) To do titration problems, we first must have the chemical equation:

$$HCl + NaOH \rightarrow H_2O + NaCl$$

The endpoint means that just enough acid has been added to completely react with the base.

$$\text{moles HCl} = 5.0 \frac{\text{moles}}{\cancel{\text{L}}} \times 0.0560 \cancel{\text{L}} = 0.28 \text{ moles HCl}$$

$$\frac{0.28 \cancel{\text{moles HCl}}}{1} \times \frac{1 \text{ mole NaOH}}{1 \cancel{\text{moles HCl}}} = 0.28 \text{ moles NaOH}$$

$$\text{concentration} = \frac{\# \text{ moles}}{\# \text{ liters}} = \frac{0.28 \text{ moles}}{0.1200 \text{ L}} = \underline{2.3 \text{ M}}$$

15. (3 pts, one for the balanced equation, and two for the stoichiometry) To do any titration problem, we must first get the chemical equation:

$$H_2SO_4 + 2NaOH \rightarrow 2H_2O + Na_2SO_4$$

The endpoint means that just enough base has been added to completely react with the acid.

$$\text{moles NaOH} = 3.0 \frac{\text{moles}}{\cancel{\text{L}}} \times 0.0034\ \cancel{\text{L}} = 0.010\ \text{moles NaOH}$$

$$\frac{0.010\ \cancel{\text{moles NaOH}}}{1} \times \frac{1\ \text{mole}\ H_2SO_4}{2\ \cancel{\text{moles NaOH}}} = 0.0050\ \text{moles}\ H_2SO_4$$

$$\text{concentration} = \frac{\#\ \text{moles}}{\#\ \text{liters}} = \frac{0.0050\ \text{moles}}{0.3000\ \text{L}} = \underline{0.017\text{M}}$$

Total possible points: 26

SOLUTIONS TO THE MODULE #11 TEST

1. (1 pt) Solubility is the maximum amount of solute that can dissolve in a given amount of solvent. It is usually expressed in grams of solute per 100 grams of solvent.

2. (1 pt) It is most likely a liquid. A liquid's solubility in another liquid is not affected by temperature or pressure.

3. (1 pt) It is, most likely, a gas. A gas's solubility in liquid decreases with increasing temperature.

4. (1 pt) It dissolves exothermically. Remember, exothermic processes *release* heat, making their surroundings hotter.

5. (1 pt) Heat up the alcohol. This will drive most of the gas out.

6. (1 pt) Solids increase in solubility with increasing temperature.

7. (1 pt) The solute that will raise the boiling point the most is the one that splits into the most ions when dissolved. This would be Na_3PO_4, which splits into four ions.

8. (1 pt) He has divided by the mass of the *solution.* Molality is calculated by dividing by the mass of the *solvent*.

9. (3 pts, one for each step) This is just a stoichiometry problem:

$$\frac{350.0\ \cancel{\text{g CuI}_2}}{1} \times \frac{1\text{ mole CuI}_2}{317.3\ \cancel{\text{g CuI}_2}} = 1.103\text{ moles CuI}_2$$

$$\frac{1.103\ \cancel{\text{moles CuI}_2}}{1} \times \frac{2\text{ moles NaI}}{1\ \cancel{\text{mole CuI}_2}} = 2.206\text{ moles NaI}$$

$$\frac{2.206\ \cancel{\text{moles NaI}}}{1} \times \frac{1\text{ L NaI}}{3.3\ \cancel{\text{moles NaI}}} = 0.67\text{ L} = \underline{670\text{ mL}}$$

10. (3 pts, one for each step) This is just a stoichiometry problem:

$$\frac{0.035\ \cancel{\text{L H}_2\text{O}_2}}{1} \times \frac{1.20\text{ mole H}_2\text{O}_2}{1\ \cancel{\text{L H}_2\text{O}_2}} = 0.042\text{ moles H}_2\text{O}_2$$

$$\frac{0.042\ \cancel{\text{moles H}_2\text{O}_2}}{1} \times \frac{1\text{ mole O}_2}{2\ \cancel{\text{moles H}_2\text{O}_2}} = 0.021\text{ moles O}_2$$

$$\frac{0.021\ \cancel{\text{moles O}_2}}{1} \times \frac{32.0\text{ g O}_2}{1\ \cancel{\text{mole O}_2}} = \underline{0.67\text{ g O}_2}$$

11. (1 pt)

$$m = \frac{\text{\# moles solute}}{\text{\# kg solvent}} = \frac{5.61 \text{ moles}}{2.11 \text{ kg}} = 2.66 \frac{\text{moles}}{\text{kg}} = \underline{2.66 \text{ m}}$$

12. (2 pts, one for #moles, and one for grams)

$$m = \frac{\text{\# moles solute}}{\text{\# kg solvent}}$$

$$2.13 \text{ m} = \frac{\text{\# moles}}{0.670 \text{ kg}}$$

$$\text{\# moles} = 1.43$$

$$\frac{1.43 \cancel{\text{moles KBr}}}{1} \times \frac{119.0 \text{ g KBr}}{1 \cancel{\text{mole KBr}}} = \underline{1.70 \times 10^2 \text{ g KBr}}$$

13. (2 pts, one for getting "i" correct, and one for the answer) In order to solve a freezing point depression problem, we need to know the value for "i." MgF_2 splits into an Mg^{2+} ion and two F^{1-} ions. Thus, i = 3.

$$\Delta T = -i \cdot K_f \cdot m$$

$$-3.5 \text{ °C} = -3 \cdot (1.86 \frac{\text{°C}}{\text{molal}}) \cdot m$$

$$m = \underline{0.63 \text{ m}}$$

14. (3 pts, one for "i", one for the molality, and one for the answer) Since CCl_4 is covalent, it does not split into ions when dissolved. Thus, i = 1. To use the freezing point depression equation, however, we need to calculate the molality of the solution.

$$\frac{30.0 \cancel{\text{g CCl}_4}}{1} \times \frac{1 \text{ mole CCl}_4}{154.0 \cancel{\text{g CCl}_4}} = 0.195 \text{ moles CCl}_4$$

$$m = \frac{\text{\# moles solute}}{\text{\# kg solvent}} = \frac{0.195 \text{ moles}}{0.3500 \text{ kg}} = 0.557 \text{ m}$$

Now we can use the freezing point depression equation:

$$\Delta T = -i \cdot K_f \cdot m$$

$$\Delta T = -1 \cdot (5.12 \frac{^{\circ}C}{\cancel{m}}) \cdot 0.557\cancel{m}$$

$$\Delta T = -2.85\,^{\circ}C$$

This is not the answer. This tells us how much the freezing point lowered! Thus, the new freezing point is 5.5 °C - 2.85 °C = 2.7 °C.

15. (2 pts, one for "i", and one for the answer) Since $Ca(NO_3)_2$ splits into one Ca^{2+} and two NO_3^- ions, i = 3.

$$\Delta T = i \cdot K_b \cdot m$$

$$\Delta T = 3 \cdot (3.08 \frac{^{\circ}C}{\cancel{m}}) \cdot 3.20\cancel{m}$$

$$\Delta T = 29.6\,^{\circ}C$$

This is not the answer. This tells us how much the boiling point raised! Thus, the new boiling point is 118.5 °C + 29.6 °C = 148.1 °C.

Total possible points: 24

SOLUTIONS TO THE MODULE #12 TEST

1. (1 pt) Charles' Law states that under conditions of constant pressure, the temperature and volume are linearly proportional to each other.

2. (1 pt) pound

3. (1 pt) The Kelvin temperature scale is absolute because you can never get to or below 0 K.

4. (1 pt) Gases behave ideally near STP. If the temperature is higher than STP and/or the pressure is lower, that's even more ideal. The situation in b is therefore the most ideal.

5. (1 pt) The boiling point of a liquid is the temperature at which the liquid's vapor pressure is equal to atmospheric pressure.

6. (1 pt) The second will result in a higher mole fraction of hydrogen, because the vapor pressure of water will be lowest in that experiment.

7. (1 pt) He made an extrapolation with far too little data.

8. (1 pt)

$$\frac{\cancel{P_1}V_1}{T_1} = \frac{\cancel{P_2}V_2}{T_2}$$

$$\frac{(189\text{ L})}{294\text{ K}} = \frac{(4.5 \times 10^2\text{ L})}{T_2}$$

$$T_2 = \frac{(4.5 \times 10^2\ \cancel{\text{L}}) \cdot 294\text{ K}}{189\ \cancel{\text{L}}} = \underline{7.0 \times 10^2\text{ K}} \text{ or } 430\ ^\circ\text{C}$$

9. (1 pt)

$$\frac{P_1V_1}{T_1} = \frac{P_2V_2}{T_2}$$

$$\frac{(1.02\text{ atm}) \cdot (34.6\text{ l})}{297\text{ K}} = \frac{(0.21\text{ atm}) \cdot V_2}{284\text{ K}}$$

$$V_2 = \frac{(1.02\ \cancel{\text{atm}}) \cdot (34.6\text{ L}) \cdot 284\ \cancel{\text{K}}}{(0.21\ \cancel{\text{atm}}) \cdot 297\ \cancel{\text{K}}} = \underline{1.6 \times 10^2\text{ L}}$$

10. (1 pt) $P_{total} = P_{oxygen} + P_{vapor} = 567\text{ torr} + 30.0\text{ torr} = \underline{597\text{ torr}}$

11. (2 pts, one for getting everthing into moles, and one for the answers) To calculate mole fractions, we must get everything into moles:

$$\frac{12.0\ \cancel{g\,O_2}}{1} \times \frac{1\ \text{mole}\,O_2}{32.0\ \cancel{g\,O_2}} = 0.375\ \text{moles}\,O_2$$

$$\frac{100.0\ \cancel{g\,C_2H_2}}{1} \times \frac{1\ \text{mole}\ C_2H_2}{26.0\ \cancel{g\,C_2H_2}} = 3.85\ \text{moles}\,C_2H_2$$

Now we can use the definition of mole fraction.

$$X_{O_2} = \frac{0.375\ \cancel{\text{moles}}}{4.23\ \cancel{\text{moles}}} = \underline{0.0887}$$

$$X_{C_2H_2} = \frac{3.85\ \cancel{\text{moles}}}{4.23\ \cancel{\text{moles}}} = \underline{0.910}$$

12. (1 pt) Rather than use the definition of mole fraction here, it is best to use Dalton's law.

$$X_{SO_2} = \frac{0.32\ \cancel{\text{atms}}}{1.77\ \cancel{\text{atms}}} = \underline{0.18}$$

$$X_{SO_3} = \frac{1.45\ \cancel{\text{atms}}}{1.77\ \cancel{\text{atms}}} = \underline{0.819}$$

13. (2 pts, one for converting to moles, and one for the answer) This uses the ideal gas law, but first we need to convert grams to moles:

$$\frac{345.1\ \cancel{g\,CO_2}}{1} \times \frac{1\ \text{mole}\,CO_2}{44.0\ \cancel{g\,CO_2}} = 7.84\ \text{moles}\,CO_2$$

Now we can use the ideal gas law:

$$PV = nRT$$

$$P = \frac{nRT}{V} = \frac{7.84\ \cancel{\text{moles}} \cdot 0.0821 \frac{\cancel{L} \cdot \text{atm}}{\cancel{\text{mole}} \cdot \cancel{K}} \cdot 293.2\ \cancel{K}}{32.1\ \cancel{L}} = \underline{5.88\ \text{atm}}$$

The pressure will be 5.88 atm.

14. (3 pts, one for each step) This is a stoichiometry problem, once we use the ideal gas law to calculate the moles of oxygen:

$$n = \frac{PV}{RT} = \frac{0.97\ \cancel{atm} \cdot 1.4\ \cancel{L}}{0.0821\ \frac{\cancel{L} \cdot \cancel{atm}}{mole \cdot \cancel{K}} \cdot 328\ \cancel{K}} = 0.050\ \text{moles}$$

Now that we have the moles of limiting reactant, this is a stoichiometry problem.

$$\frac{0.050\ \cancel{moles\,O_2}}{1} \times \frac{4\ \text{moles Fe}}{3\ \cancel{moles\,O_2}} = 0.067\ \text{moles Fe}$$

$$\frac{0.067\ \cancel{moles\,Fe}}{1} \times \frac{55.8\ \text{g Fe}}{1\ \cancel{mole\,Fe}} = \underline{3.7\ \text{g Fe}}$$

15. (3 pts, one for each step) This is like #14, but we need to do it backwards. We start with stoichiometry and end with the ideal gas law:

$$\frac{1.3\ \cancel{g\,Mg}}{1} \times \frac{1\ \text{mole Mg}}{24.3\ \cancel{g\,Mg}} = 0.053\ \text{moles Mg}$$

$$\frac{0.053\ \cancel{moles\,Mg}}{1} \times \frac{1\ \text{moles}\,O_2}{2\ \cancel{moles\,Mg}} = 0.027\ \text{moles}\,O_2$$

Now we use the ideal gas law to calculate pressure.

$$PV = nRT$$

$$P = \frac{nRT}{V} = \frac{0.027\ \cancel{moles} \cdot 0.0821 \frac{\cancel{L} \cdot atm}{\cancel{mole} \cdot \cancel{K}} \cdot 298\ \cancel{K}}{0.0100\ \cancel{L}} = \underline{66\ \text{atm}}$$

Total possible points: 21

SOLUTIONS TO THE MODULE #13 TEST

1. (1 pt) Exothermic Reactions.

2. (1 pt) Hess's Law states that enthalpy is a state function and is therefore independent of path. The equation is: $\Delta H = \Sigma \Delta H_f$ (products) - $\Sigma\ \Delta H_f$ (reactants).

3. (2 pts, take off one point if diagram III is listed) Diagrams I and II, because the products have more energy than the reactants.

4. (1 pt) Diagram III, because the hump is the smallest.

5. (1 pt) The answer may vary a bit, depending on how you read the graph. The ending energy (400 kJ) minus the beginning energy (100 kJ) is the ΔH, so the answer should be around 300 kJ.

6. (1 pt) The 50.0 g sample has more entropy. The more molecules there are to keep track of, the more disordered the sample.

7. (1 pt) Raise the temperature. Remember, $\Delta G = \Delta H - T \cdot \Delta S$, and ΔG must be negative to make a reaction spontaneous. For endothermic reactions ΔH is positive. If ΔS is also positive, then the only way to make ΔG negative is to have a large value for T. That way, a large number will be subtracted from the ΔH. If the number is large enough, ΔG will be negative.

8. (2 pts, one for the Lewis structures, and one for the answer) The ideal way to calculate ΔH is with Hess' Law. Unfortunately, Table 13.2 does not have methane listed. Thus, we cannot use Hess' law. We must therefore use bond energies.

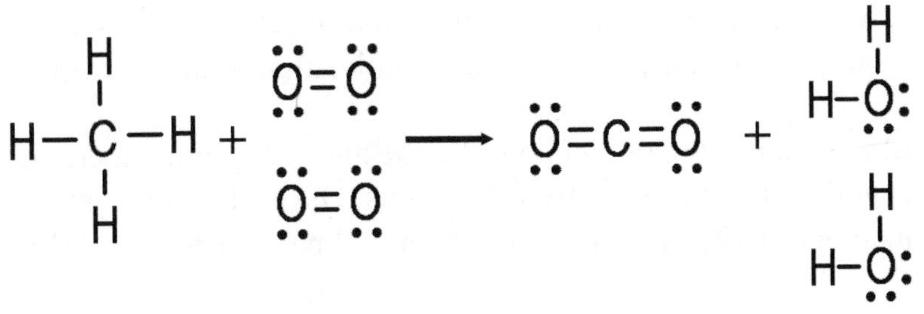

$$\Delta H = (4\ \cancel{\text{moles}}) \times (411 \frac{\text{kJ}}{\cancel{\text{mole}}}) + (2\ \cancel{\text{moles}}) \times (494 \frac{\text{kJ}}{\cancel{\text{mole}}}) - (2\ \cancel{\text{moles}}) \times (799 \frac{\text{kJ}}{\cancel{\text{mole}}}) - (4\ \cancel{\text{moles}}) \times (459 \frac{\text{kJ}}{\cancel{\text{mole}}})$$

$$\Delta H = \underline{-802\ \text{kJ}}$$

9. (2 pts, one for the balanced equation, and one for the answer) The combustion of CH_3OH (l) is given by:

$$2CH_3OH\ (l) + 3O_2\ (g) \rightarrow 2CO_2\ (g) + 4H_2O\ (g)$$

Since all of the compounds in this equation are in Table 13.2, we can use Hess's Law. Remember, ΔH_f of O_2 (g) is zero because it is an element in its elemental form!

$$\Delta H^\circ = (2\ \cancel{\text{moles}}) \times (-394 \frac{\text{kJ}}{\cancel{\text{mole}}}) + (4\ \cancel{\text{moles}}) \times (-242 \frac{\text{kJ}}{\cancel{\text{mole}}}) - (2\ \cancel{\text{moles}}) \times (-239 \frac{\text{kJ}}{\cancel{\text{mole}}})$$

$$\Delta H^\circ = \underline{-1{,}278\ \text{kJ}}$$

10. (2 pts, one for each step) The answer to #9 tells us that this reaction has 1,278 kJ as a product. Thus, we can use it in stoichiometry:

$$\frac{245.0\ \cancel{\text{g}\,CH_3OH}}{1} \times \frac{1\ \text{mole}\ CH_3OH}{32.0\ \cancel{\text{g}\,CH_3OH}} = 7.66\ \text{moles}\ CH_3OH$$

$$\frac{7.66\ \cancel{\text{moles}\,CH_3OH}}{1} \times \frac{1{,}278\ \text{kJ}}{2\ \cancel{\text{moles}\,CH_3OH}} = \underline{4{,}890\ \text{kJ}}$$

If the answer to #9 was wrong, give the student full credit as long as he or she used it properly in this problem.

11. (1 pt) The bond energies method is not as accurate as Hess's Law because it does not take the phases of the substances into account.

12. a. (1 pt) positive, because there are more gas molecules on the products side
  b. (1 pt) negative, because there are more gas molecules on the reactants side
  c. (1 pt) negative, because there are more aqueous molecules on the reactants side

13. (2 pts, one for setting up the equation, and one for the answer) Remember, only enthalpies of formation of elements in their natural form are zero. Thus, O_2 (g) does have an absolute entropy. To get it, we will use Equation (13.12). The only unknown will be the ΔS of O_2, so we can solve for it:

$$209.6 \frac{\text{J}}{\text{K}} = (2\ \cancel{\text{moles}}) \times (189 \frac{\text{J}}{\cancel{\text{mole}} \cdot \text{K}}) + (2\ \cancel{\text{moles}}) \times (214 \frac{\text{J}}{\cancel{\text{mole}} \cdot \text{K}}) - (1\ \cancel{\text{mole}}) \times 219 \frac{\text{J}}{\cancel{\text{mole}} \cdot \text{K}} - (3\ \text{moles}) \times (\Delta S_{O_2})$$

$$\Delta S_{O_2} = \frac{2 \times (189 \frac{\text{J}}{\text{K}}) + 2 \times (214 \frac{\text{J}}{\text{K}}) - 219 \frac{\text{J}}{\text{K}} - 209.6 \frac{\text{J}}{\text{K}}}{3\ \text{moles}} = \underline{126 \frac{\text{J}}{\text{mole} \cdot \text{K}}}$$

14. (1 pt) This is also a Hess' Law problem. Remember, ΔG_f of an element in its natural form is zero!

$$\Delta G^\circ = (12 \cancel{\text{moles}}) \times (-394 \frac{\text{kJ}}{\cancel{\text{mole}}}) + (6 \cancel{\text{moles}}) \times (-229 \frac{\text{kJ}}{\cancel{\text{mole}}}) - (2 \cancel{\text{moles}}) \times (125 \frac{\text{kJ}}{\cancel{\text{mole}}}) = \underline{-6{,}352 \text{ kJ}}$$

15. (1 pt) To be spontaneous $\Delta G < 0$. It if it spontaneous for all temperatures less than 200 K, then ΔG must equal 0 at 200 K. Thus, $\Delta H - T \cdot \Delta S = 0$

$$-120.0 \frac{\text{kJ}}{\text{mole}} - 200.0\,\text{K} \cdot (\Delta S) = 0$$

$$\Delta S = \frac{-120.0 \frac{\text{kJ}}{\text{mole}}}{200.0\,\text{K}} = \underline{-0.6000 \frac{\text{kJ}}{\text{mole} \cdot \text{K}}} \text{ or } -600.0 \frac{\text{J}}{\text{mole} \cdot \text{K}}$$

Total possible points: 22

SOLUTIONS TO THE MODULE #14 TEST

1. (1 pt) When the NaI solution is boiled, it is concentrated, thus, the new reaction rate will be larger.

2. (1 pt) Increasing temperature increases the motion of the reactant molecules. This makes collisions more likely to happen, increasing the reaction rate.

3. (1 pt) Reaction rate is unaffected by the concentration of that reactant.

4. (1 pt) The rate constant depends on temperature.

5. (1 pt) Their rates should be the same.

6. (2 pts, one for each definition) Homogeneous catalysts are catalysts that have the same phase as the reactants. Heterogeneous catalysts have a different phase than the reactants.

7. (1 pt) HCl is not a catalyst here because it gets used up. Catalysts are never used up in the reaction.

8. (3 pts, one for each order, and one for k) The rate equation will have the form:

$$R = k\,[C_3H_6Cl_2]^x[Br^-]^y$$

In trials 1&2, the concentration of Br^- stayed the same but the concentration of $C_3H_6Cl_2$ doubled. This resulted in a doubling of the reaction rate, thus x=1. In trials 2&3, the concentration of $C_3H_6Cl_2$ stayed the same but the concentration of Br^- doubled. This resulted in no change in the rate, thus y = 0.

$$R = k[C_3H_6Cl_2]$$

$$0.116\,\frac{M}{s} = k\cdot(0.080\ M)$$

$$k = \frac{0.116\,\frac{\not{M}}{s}}{(0.080\ \not{M})} = 1.5\,\frac{1}{s}$$

The rate equation is $R = (1.5\,\frac{1}{s})\cdot[C_3H_6Cl_2]$.

9. (3 pts, one for each order, and one for k) The rate equation will have the form:

$$R = k\,[NO]^x[I_2]^y$$

In trials 1&2, the concentration of NO stayed the same but the concentration of I_2 doubled. This resulted in a doubling of the reaction rate, thus y = 1. In trials 2&3, the concentration of I_2 stayed the same but the concentration of NO doubled. This resulted in a quadrupling of the reaction rate, thus x=2.

$$R = k[NO]^2[I_2]$$

$$1.25\,\frac{M}{s} = k \cdot (1.5\ M)^2 \cdot (1.0\ M)$$

$$k = \frac{1.25\,\frac{\cancel{M}}{s}}{(1.5\ M)^2 \cdot (1.0\ \cancel{M})} = 0.56\,\frac{1}{M^2 \cdot s}$$

The rate equation is $R = (0.56\,\frac{1}{M^2 \cdot s}) \cdot [NO]^2[I_2]$.

10. (1 pt) You do not need the temperature to solve this problem. That was put there to fool you. This is simply an application of the rate equation. The problem tells us:

$$R = (0.0123\,\frac{1}{M \cdot s}) \cdot [CH_3OH] \cdot [KF]$$

$$R = (0.0123\,\frac{1}{\cancel{M} \cdot s}) \cdot (0.59\ \cancel{M}) \cdot (1.22\ M) = 0.0089\,\frac{M}{s}$$

11. (1 pt) The overall order is the sum of the individual orders, which is 5.

12. (1 pt) The rate should double for every 10 °C increase. This rate, then, should go up by 2x2x2 = 8. The resulting rate is 0.0361 M/s.

13. (1 pt) The rate will decrease by a factor of 2 for every 10 °C the temperature is lowered. To slow down by a factor of 4, the temperature must be dropped 20 °C to 5 °C.

14. (3 pts, one for each energy diagram). The only thing that changes from diagram to diagram is the height of the hump. The faster the rate, the lower the height of the hump:

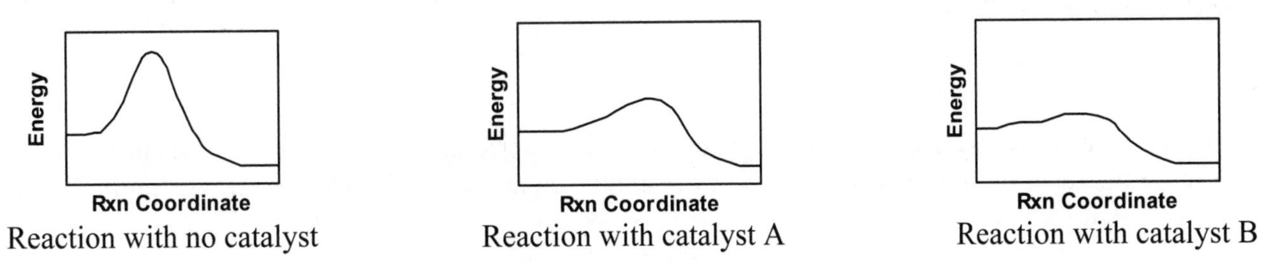

Reaction with no catalyst — Reaction with catalyst A — Reaction with catalyst B

15. (1 pt) Mn^{2+}: it is not used up in the reaction. It is consumed in step 1 but produced in step 3.

Total possible points: 22

SOLUTIONS TO THE MODULE #15 TEST

1. (1 pt) Equilibrium occurs when the rates of the forward and reverse reactions equal. This occurs at about 18 seconds, according to the graph. Your number can be slightly different than mine, since you are reading it from a graph.

2. (1 pt) When a stress (such as a change in concentration, pressure, or temperature) is applied to an equilibrium, the reaction will shift in a way that relieves the stress and restores equilibrium.

3. (1 pt) The acidity or basicity of a solution.

4. (1 pt) $HNO_3\ (aq) \leftrightarrows H^+\ (aq) + NO_3^-\ (aq)$

5. (1 pt) When an equilibrium is stressed, one of the rates (forward or reverse) changes dramatically. On the graph, this seems to occur at about 29 seconds. Your number can be slightly different than mine, since you are reading it from a graph.

6. (4 pts) Strongest Acid: E Weakest Acid: D Strongest Base: B Weakest Base: C

7. (2 pts) The higher the pH, the weaker the acid. The weaker the acid, the smaller the ionization constant. Thus, high pH means low ionization constant. The order, then, is $A < B < D < E < C$

8. (1 pt) You take the concentrations and equilibrium and plug them into the equation:

$$K = \frac{[NO]^2 \cdot [Br_2]}{[NOBr]^2} = \frac{(0.010\ \cancel{M})^2 \cdot (0.0050\ M)}{(0.10\ \cancel{M})^2} = 5.0 \times 10^{-5}\ M$$

9. (1 pt) Since $BaSO_3$ and BaO are both solids, they are ignored in the equation for the equilibrium constant. Thus, $K = [SO_2]$. This means that at equilibrium, the concentration of SO_2 is the same as K, which is 0.345 M.

10. Reactions can be written with a single arrow forward if $K \gg 1$ and with a single arrow in reverse if $K \ll 1$.

a. (1 pt) $2NO_2\ (g) + Cl_2\ (g) \rightarrow 2NOCl\ (g)$

c. (1 pt) $2NO_2 \rightarrow N_2O_4$

Subtract 1 point if reaction (b) is listed.

11. (2 pts, one for working out the equation, and one for determining the direction of the shift) To test and see if the reaction is at equilibrium, we evaluate the equation for K. Remember, we ignore liquids!

$$K = \frac{[H_2SO_3]}{[SO_2]} = \frac{0.23\ \cancel{M}}{0.35\cancel{M}} = 0.66$$

At this point, K is too large compared to its correct value. To reduce the number, we must get less products and more reactants. Thus, the reaction must shift towards the reactants in order to reach equilibrium.

12. a. (1 pt) Nothing will happen, because solids do not affect equilibrium.
    b. (1 pt) The amount of MgO will increase, because the reaction will shift towards the products.
    c. (1 pt) Nothing will happen, because liquids are ignored.

13. a. (1 pt) The concentration of O_2 will decrease, because the ΔH tells us that energy is a reactant. When you raise temperature, you are therefore adding a reactant, shifting the equilibrium to the products side.
    b. (1 pt) The concentration of NO_2 will increase, because raising pressure shifts the equilibrium to the side with the least molecules of gas.
    c. (1 pt) The concentration of N_2 will increase, because lowering pressure shifts the equilibrium to the side with the most molecules of gas.

14. (1 pt) The acid ionization reaction is:

$$HClO_2\ (aq) \leftrightarrows H^+\ (aq) + ClO_2^-\ (aq)$$

The acid ionization constant is the equilibrium constant for that reaction:

$$K = \frac{[ClO_2^-][H^+]}{[HClO_2]}$$

15. (1 pt) The base ionization reaction is:

$$CH_5P\ (aq) + H_2O\ (l) \leftrightarrows OH^-\ (aq) + CH_6P^+\ (aq)$$

The base ionization constant is the equilibrium constant for that reaction:

$$K = \frac{[CH_6P^+][OH^-]}{[CH_5P]}$$

Total possible points: 25

SOLUTIONS TO THE MODULE #16 TEST

1. (1 pt) Oxidation numbers are calculated assuming that the most electronegative atoms in a molecule actually take the shared electrons away from the less electronegative atoms. This never actually happens in covalent compounds, so the charges never really exist.

2. (1 pt) The oxidation number will most likely decrease (become more negative). If the other atoms are less electronegative, they will pull away less electrons, allowing the atom in question to get more. More electrons means a more negative charge.

3. (1 pt) Reduction must always accompany oxidation because electrons cannot simply appear or disappear. If one substance is going to lose electrons, another substance must be there to gain them.

4. (1 pt) The anode is the negative side of the battery (a source of anions), and the cathode is the positive side of the battery (a source of cations). Thus, electrons will flow from the anode to the cathode.

5. (1 pt) The alkaline battery is not rechargeable because its inner construction degrades.

6. (1 pt) The lead-acid battery also uses PbO_2, which is lead(IV) oxide.

7. a. (1 pt) Mn: +4, O: -2
   b. (1 pt) P: +5, O: -2
   c. (1 pt) S: +4, O: -2
   d. (1 pt) Mg: +2, Br: -1
   e. (1 pt) K: +1, Cl: +5, O: -2
   f. (1 pt) P: +4, F: -1
   g. (1 pt) Pt:+2, Cl: -1
   h. (1 pt) N: 0

8. (2 pts) oxidized, 2 electrons

9. (2 pts) reduced, 2 electrons

10. (2 pts) reduced, 4 electrons

11. (2 pts) oxidized, 3 electrons

12. a. (1 pt) REDOX: O is reduced, Cl is oxidized
    b. (1 pt) REDOX: Cu is oxidized, N is reduced
    c. (1 pt) NOT REDOX
    d. (1 pt) REDOX: S is oxidized, I is reduced
    e. (1 pt) NOT REDOX

13. (2 pts)

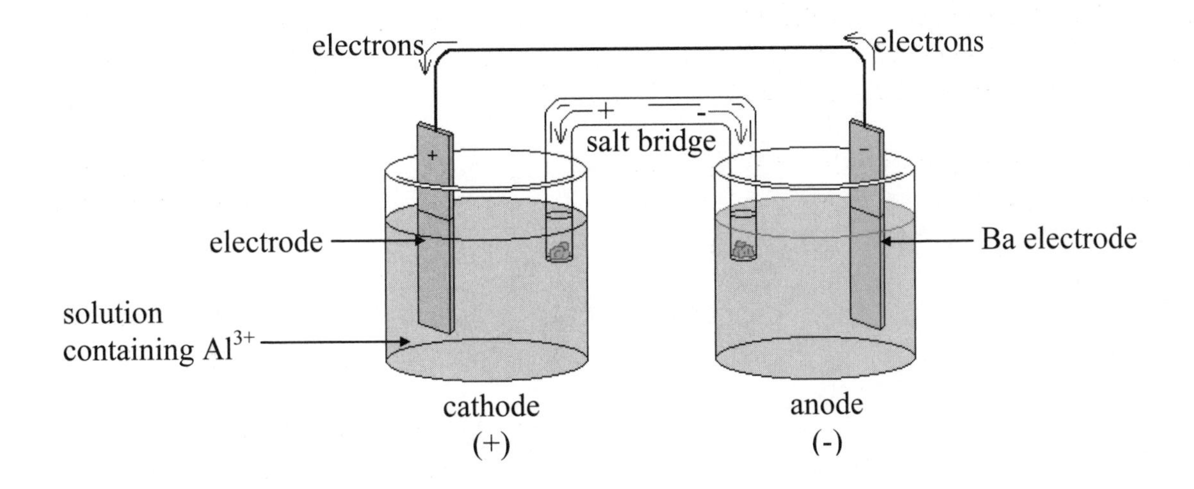

14. (2 pts)

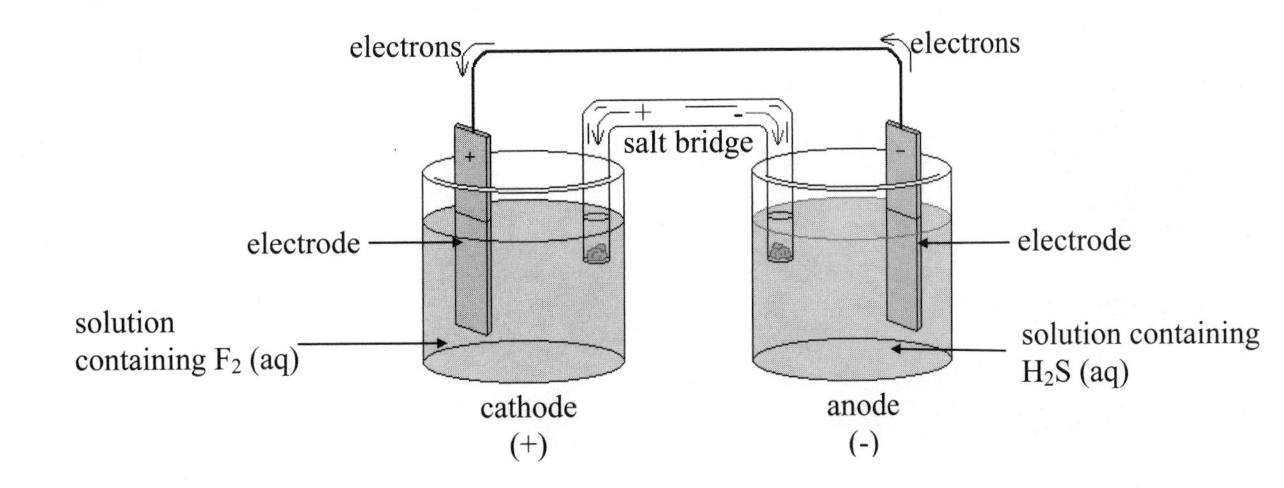

Total possible points: 31

QUARTERLY TEST #1

1. Which is longer, a string that measures 0.25 km or a string measuring 101 mm?

2. How long is the black bar in the picture below?

3. Two students measure the length of a 2.1-meter object. The first student measures the length to be 2.8123 m. The second measure the length to be 2 m. Which student was more precise? Which student was more accurate?

4. Convert 34.5 kL into L.

5. Convert 0.001400 into scientific notation.

6. Convert 0.00045 kg to mg.

7. Gold has a density of 19.3 grams per mL. If a gold nugget has a volume of 5.60 L, what is its mass?

8. A young girl puts a leash on her dog, but she can't get the dog to move. She pulls and pulls and pulls but the dog will not budge. Did the girl do any work?

9. Classify each of the following as having either potential or kinetic energy, or both:

a. A pile of fire wood
b. Light
c. Wind
d. A gallon of gasoline

10. Examine following the heating curve for an unknown substance:

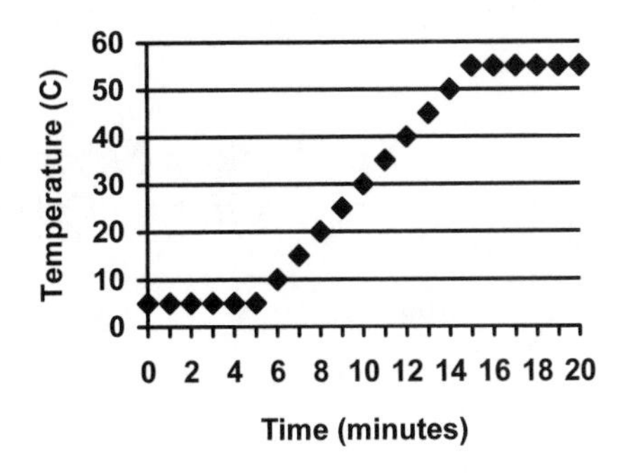

At what temperature does this substance melt?

11. How many Joules does it take to heat up 11.5 kg of glass from 25 °C to 45 °C? (The specific heat capacity of glass is 0.837 $\frac{J}{g \cdot {}^{o}C}$.)

12. A 25.0-g piece of copper (specific heat capacity = 0.385$\frac{J}{g \cdot {}^{o}C}$) at room temperature (25°C) loses 856.7 J of heat. What is its new temperature?

13. A calorimeter is filled with 150.0 g of water at 22.1 °C. A 50.0 g sample of a metal at 100.0 °C is dropped in this calorimeter and causes the temperature to increase a total of 3.4 °C. What is the specific heat of the metal? Ignore the heat absorbed by the calorimeter.

14. Write the four assumptions of Dalton's atomic theory.

15. How can you experimentally determine whether a compound is ionic or covalent?

16. How can you determine from the periodic chart whether a compound is ionic or covalent?

17. Which of the following molecules are covalent?

a. CO_2 b. $CaBr_2$ c. KCl d. C_3H_8

18. In making magnesium bromide, 100.0 g of magnesium plus 100.0 g of bromine makes 115.2 g of product along with some left over magnesium. How much magnesium and bromine should be added together in order to make 1.000 kg of magnesium bromide without any left overs?

19. For the following compounds, if the name is given, write its formula. If the formula is given, write its name.

a. diphosphorus hexaoxide b. PH_3 c. nitrogen dioxide d. CaS

20. If a liquid goes through a phase change and all you know is that the molecules slowed down and moved closer together, what phase did the liquid turn in to?

21. Classify the following as an element, compound, homogeneous mixture, or heterogeneous mixture.

a. A fruitcake
b. A bottle of CH_4O
c. A pile of carbon powder
d. Sugar dissolved in water

22. Classify the following as physical or chemical changes:

a. A priceless vase is shattered.
b. Coal burns in a furnace.
c. Water freezes.
d. Iron rusts.

23. Balance the following equation:

$$HNO_3\ (aq) + Mg\ (s) \rightarrow MgN_2O_6\ (aq) + H_2\ (g)$$

24. Balance the following equation:

$$C_{10}H_{22} + O_2 \rightarrow CO_2 + H_2O$$

25. Balance the following equation:

$$CO_2\ (g) + H_2O\ (l) \rightarrow C_6H_{12}O_6\ (s) + O_2\ (g)$$

QUARTERLY TEST #2

(1.00 amu = 1.66×10^{-24} g)

1. Classify the following reactions as decomposition, formation, complete combustion, or none of these:

   a. C_6H_{12} (s) + $9O_2$ (g) → $6CO_2$ (g) + $6H_2O$ (g)

   b. N_2 (g) + $3H_2$ (g) → $2NH_3$ (g)

   c. 2KCl (s) + $MgBr_2$ (aq) → $MgCl_2$ (s) + 2KBr (aq)

   d. $2H_3PO_4$ (s) → $3H_2$ (g) + 2P (s) + $4O_2$ (g)

2. What is the difference between complete combustion and incomplete combustion?

3. Write a balanced chemical equation for the for the formation of $NaHSO_4$.

4. Write a balanced chemical equation for the complete combustion of octane, C_8H_{18}.

5. What is the mass of an $MgCO_3$ molecule in kg? (1.00 amu = 1.66×10^{-24} g)

6. How many moles of K_2CO_3 are in a 150.0 gram sample of the compound?

7. What is the mass of an $MgBr_2$ sample if it contains 0.342 moles of the compound?

8. In the following reaction:

$$2KCl\ (s) + MgBr_2\ (aq) \rightarrow MgCl_2\ (s) + 2KBr\ (aq)$$

2 moles of KCl are reacted with 2 moles of $MgBr_2$. What is the limiting reactant?

9. A chemist wants to perform the following reaction:

$$MgCO_3\ (s) + 2HCl\ (g) \rightarrow MgCl_2\ (aq) + CO_2\ (g) + H_2O\ (l)$$

Which substances can she use Gay-Lussac's law to relate to one another?

10. Which of the following are empirical formulas?

    a. $C_{18}H_{36}O_{18}$ b. $Na_2S_2O_3$ c. CaN_2O_6

11. Freon, a very useful refrigerant, is produced in the following reaction:

$$3CCl_4\ (g) + 2SbF_3\ (s) \rightarrow \underset{\text{(Freon)}}{3CCl_2F_2\ (g)} + 2SbCl_3\ (s)$$

Suppose a chemist wanted to make 115.0 liters of Freon using excess antimony trifluoride. How many liters of carbon tetrachloride would the chemist need?

12. To make a phosphorus fertilizer, agricultural companies use the following reaction:

$$Ca_3P_2O_8 + 2H_2SO_4 + 4H_2O \rightarrow \underset{\text{(fertilizer)}}{CaH_4P_2O_8} + 2CaH_4SO_6$$

If 150.0 grams of $Ca_3P_2O_8$ are reacted with excess H_2SO_4 and H_2O, how many grams of fertilizer can be made?

13. In the following reaction:

$$Mg\ (s) + 2H_2O\ (l) \rightarrow Mg(OH)_2\ (s) + H_2\ (g)$$

If you have 4.1×10^3 grams of Mg, how many grams of water must be used to have all of the Mg react to form $Mg(OH)_2$?

14. A compound has an empirical formula of CH. If its molecular mass is 78.1 amu, what is its molecular formula?

15. An unknown compound was decomposed into 63.2 g carbon, 5.26 g hydrogen, and 41.6 g oxygen. What is its empirical formula?

16. If a substance has 5 positive charges and 4 negative charges, is it positively charged, negatively charged, or neutral?

17. If you have an yellow light bulb and a red one, which emits waves with the largest wavelength? Which emits light of higher frequency? Which emits the higher energy light?

18. Give the number of protons, electrons, and neutrons in the following atoms:

a. ^{39}K b. ^{20}Ne c. ^{52}Mn d. ^{131}I

19. Give the full electron configurations of the following atoms:

a. Fe b. P c. K

20. Give the abbreviated electron configurations for the following atoms:

a. Cl b. Nb c. Cs

21. What is the fundamental difference between metals and nonmetals?

22. Draw the Lewis structures for the following atoms:

a. Be b. C c. Cl

23. Give the chemical formulas for the following compounds:

a. aluminum sulfide b. potassium nitride c. magnesium sulfide d. chromium(III) oxide

24. What is the Lewis structure for NH_3?

25. What is the Lewis structure for H_2CO?

QUARTERLY TEST #3

1. Give the chemical formulas for the following compounds:

a. potassium carbonate b. calcium nitrate c. aluminum sulfate d. calcium chromate

2. Name the following compounds

a. $Ca_3(PO_4)_2$ b. $Ca(NO_3)_2$ c. NH_4Cl d. K_2CO_3

3. Determine the shape of a CO molecule. Give its bond angle and draw a picture of it.

4. Determine the shape of a CCl_4 molecule. Give its bond angle and draw a picture of it.

5. Determine the shape of a NF_3 molecule. Give its bond angle and draw a picture of it.

6. Determine the shape of a CS_2 molecule. Give its bond angle and draw a picture of it.

7. Identify any of the molecules in Problems 3-6 that are purely covalent.

8. Identify the acid in the following reaction:

$$C_2H_6O + CH_5N \rightarrow C_2H_5O^- + CH_6N^+$$

9. When one molecule of $Al_2(SO_4)_3$ dissolves in water, what ions does it form and how many of each ion are present?

10. Give the balanced chemical equation for the reaction that occurs between nitric acid (HNO_3) and magnesium hydroxide.

11. Given that PH_3 is a base, determine the reaction that occurs between HNO_3 and PH_3 and give the balanced chemical equation.

12. A chemist makes a "stock" solution of KOH by dissolving 1050.0 grams in enough water to make 1.50 liters of solution. If, later on, the chemist wants to use this stock solution to make 200.0 mL of 0.10 M KOH, what would the chemist need to do?

13. A chemist needs to know the concentration of some KOH that is in the laboratory. To find this out, the chemist titrates a 50.0 mL sample of the solution with 2.5 M HCl. If it takes 14.7 mL of the HCl to reach the titration endpoint, what is the concentration of the KOH solution?

14. What two things increase the solubility of a gas in a liquid?

15. If you wanted to protect water from freezing, which compound would accomplish this best: $Al(NO_3)_3$, $Ca(NO_3)_2$, or KCl? Assume that the molality of the solution is the same in each case.

16. The following reaction is performed in a lab:

$$3Na_2SO_4\ (aq) + 2Al(NO_3)_3\ (aq) \rightarrow Al_2(SO_4)_3\ (s) + 6NaNO_3\ (aq)$$

If 191 mL of 1.25 M aluminum nitrate is added to an excess of sodium sulfate, how many grams of aluminum sulfate will be produced?

17. If you want to lower water's freezing point 11 °C by adding Na_2CO_3, what must be the molality of the salt solution? (K_f for water is 1.86 $\frac{°C}{m}$)

18. How many grams of Na_2CO_3 would you have to add to 500.0 g of water to make a solution with the molality you found in Problem #17?

19. What is the boiling point of a solution made by mixing 100.0 g NaCl with 1100.0 grams of water? (K_b for water is 0.512 $\frac{°C}{m}$)

20. What are the properties of an ideal gas?

21. Under what conditions do gases behave ideally?

22. A chemist collects 156 mL of carbon dioxide gas at 25 °C and 790 torr. What would the volume of the carbon dioxide be at STP?

23. A mixture of gases has a pressure of 1.0 atm. If the mixture is analyzed and found to contain 1.2 g of N_2 and 1.5 g of O_2, what is the partial pressures of each gas in the mixture?

24. What is the volume of 15.0 grams of carbon dioxide gas at STP?

25. A chemist performs the following reaction:

$$7H_2O_2\ (aq) + N_2H_4\ (g) \rightarrow 2HNO_3\ (g) + 8H_2O\ (g)$$

If the chemist starts with 110.0 grams of H_2O_2 and an excess of N_2H_4, what volume of water vapor will be produced at a temperature of 341 °C and a pressure of 2.1 atm?

QUARTERLY TEST #4

1. Which of the following substances will have a ΔH_f of zero?

$$KOH\ (aq),\ N_2\ (g),\ Cl^-\ (aq),\ O_2\ (l)$$

2. Draw an energy diagram for a reaction that starts with reactants whose energy is 20.0 kJ and has a ΔH of 20.0 kJ. The activation energy for the reaction is 35.0 kJ.

3. An energy diagram of the reaction described in problem #2 is drawn, but the diagram is drawn taking into account the effect of a catalyst. What would be different between this energy diagram and the one you drew for problem #2?

4. Is it possible for a reaction to occur if the reaction results in a decrease in the entropy of the chemicals involved in the reaction?

5. What is the sign of ΔS for the following reaction?

$$NaHCO_3\ (aq) + HCl\ (aq) \rightarrow NaCl\ (aq) + H_2O\ (l) + CO_2\ (g)$$

6. If the ΔH of a certain reaction is 423 kJ/mole and the ΔS is 194 $\frac{\text{Joules}}{\text{mole} \cdot \text{K}}$, what is the temperature range for which this reaction is spontaneous?

7. A chemistry book lists the rate constant for a reaction as 186.1 $\frac{1}{M^3 \cdot s}$. What is the overall order of the reaction?

8. The order of a chemical reaction with respect to one of its reactants is zero. If you double the concentration of that reactant, what happens to the rate?

9. A chemist does a reaction rate analysis on the following reaction:

$$2CO\ (g) + O_2\ (g) \rightarrow 2CO_2\ (g)$$

She collects the following data:

Trial	Initial Concentration of CO (M)	Initial Concentration of O_2 (M)	Instantaneous Reaction Rate (M/s)
1	0.150	0.150	0.113
2	0.300	0.150	0.226
3	0.300	0.300	0.904

What is the rate equation for this reaction?

10. A chemist runs a chemical reaction at 25 °C and decides that it proceeds far too repidly. As a result, he decides that the reaction rate must be decreased by a factor of 4. At what temperature should the chemist run the reaction to achieve this goal?

11. In the reaction mechanism below, indicate what substance is acting like a catalyst.

Step 1: H_2O_2 (aq) + I^- (aq) $\rightarrow$ H_2O (l) + IO^- (aq)
Step 2: H_2O_2 (aq) + IO^- (aq) $\rightarrow$ H_2O (l) + I^- (aq)

12. Is the catalyst in problem #11 a heterogeneous or a homogeneous catalyst?

13. What pH is considered neutral?

14. Three solutions have the following pH:

Solution A: pH = 11
Solution B: pH = 5
Solution C: pH = 2

Which is (are) the acidic solution(s)?

15. Three acids solutions have the following pH:

Solution A: pH = 2
Solution B: pH = 5
Solution C: pH = 3

Which solution is made with the acid that has the *smallest* ionization constant?

16. A chemist is studying the following equilibrium:

$$2Pb(NO_3)_2\text{ (s)} \leftrightarrows 2PbO\text{ (s)} + 4NO_2\text{ (g)} + O_2\text{ (g)}$$

He starts out with 10 g of $Pb(NO_3)_2$ and, at equilibrium, has 2.02 g of PbO. The concentrations of NO_2 and O_2 at equilibrium are 0.25 M and 0.019 M, respectively. What is the value of the equilibrium constant?

17. The following reaction:

$$2SO_3\text{ (g)} \leftrightarrows 2SO_2\text{ (g)} + O_2\text{ (g)}$$

has an equilibrium constant equal to 0.23 M. If the following concentrations are present: $[SO_2]$ =0.480 M, $[O_2]$ = 0.561 M, $[SO_3]$ = 0.220 M, is the reaction at equilibrium? If not, which way must it shift to reach equilibrium?

18. Consider the following reaction that has reached equilibrium:

$$NH_3\text{ (aq)} + C_2H_4O_2\text{ (aq)} \leftrightarrows NH_4^+\text{ (aq)} + C_2H_3O_2^-\text{ (aq)} \qquad \Delta H = -102.1\text{ kJ}$$

a. What will happen to the concentration of NH_4^+ (aq) if the temperature is raised?
b. What will happen to the concentration of NH_4^+ (aq) if the concentration of NH_3 is lowered?

19. What is the equation for the acid ionization constant of HI?

20. Give the oxidation numbers of all atoms in $NaMnO_4$.

21. An atom changes its oxidation number from -2 to -5. Is it oxidized or reduced? How many electrons did it take to do this?

22. An atom changes its oxidation number from 0 to +2. Is it oxidized or reduced? How many electrons did it take to do this?

23. Which of the following is a redox reaction?

a. H_2CO_3 (aq) + $2NH_3$ (aq) → $2NH_4^+$ (aq) + CO_3^{2-} (aq)
b. $2VO_3^-$ (aq) + Zn (s) + $8H^+$ (aq) → $2VO^{2+}$ (aq) + Zn^{2+} (aq) + $4H_2O$ (l)
d. $Zn(NO_3)_2$ (aq) + 2NaOH (aq) → $Zn(OH)_2$ (s) + $2NaNO_3$ (aq)

24. For the redox reaction in problem #23, which substance is oxidized? Which is reduced?

25. A Galvanic cell runs on the following reaction:

$$Co\ (s) + Cu^{2+}\ (aq) \rightarrow Co^{2+}\ (aq) + Cu\ (s)$$

Draw a diagram for this Galvanic cell, labeling the electron flow, the anode and cathode, and the positive and negative sides of the Galvanic cell.

SOLUTIONS TO QUARTERLY TEST #1

1. (1 pt) Since a km is equal to 1,000 m and a mm is only equal to 0.001 m, the string measuring 0.25 km is longer.

2. (1 pt) 2.43 cm. The student's answer can range from 2.41 cm to 2.45 cm.

3. (2 pts, one for each answer) Assuming both students reported the proper number of significant figures, the first student was more precise (because there are more places to the right of the decimal), but the second student was more accurate (because it is closer to the correct value).

4. (1 pt) $\frac{34.5\ \cancel{kL}}{1} \times \frac{1{,}000\ L}{1\ \cancel{kL}} = \underline{34{,}500\ L}$

5. (1 pt) $\underline{1.400 \times 10^{-3}}$

6. (2 pts, one for each conversion step) This is a two-step conversion, since we know of no relationship between kg and mg. Thus, we must first convert kg to g and then convert g to mg. We'll do this on one line:

$$\frac{0.00045\ \cancel{kg}}{1} \times \frac{1{,}000\ \cancel{g}}{1\ \cancel{kg}} \times \frac{1\ mg}{0.001\ \cancel{g}} = \underline{450\ mg}$$

7. [2 pts, one for the conversion to L and one for using Equation (1.1)] We will use Equation (1.1), but we have to convert kg to g and rearrange the equation to solve for mass:

$$\frac{5.60\ \cancel{L}}{1} \times \frac{1\ mL}{0.001\ \cancel{L}} = 5.60 \times 10^{3}\ mL$$

$$m = \rho \cdot V$$

$$m = 19.3\ \frac{g}{mL} \cdot 5.60 \times 10^{3}\ mL = \underline{1.08 \times 10^{5}\ g}$$

8. (1 pt) The girl did no work. Motion must occur for work to be done.

9. a. (1 pt) A pile of fire wood has potential energy in it because all matter has stored energy.
   b. (1 pt) Light has no matter, but it does move, so it has kinetic energy.
   c. (1 pt) Wind is moving air, so it has kinetic energy. Air also has mass, however, so it has stored energy as well. Thus, it has both.
   d. (1 pt) The gasoline is matter, so it has potential energy.

10. (1 pt) In Experiment 2.1, we learned that the heating curve for water has two flat regions. The first occurred when ice was melting, and the second occurred when water was boiling. Based on this fact, the unknown liquid must melt at 5 °C. The answer can range from 3 °C to 7 °C.

11. [2 pts, one for the conversion and one for using Equation (2.3)] Since we have specific heat, mass, initial temperature, and final temperature, we are obviously supposed to use Equation (2.3). Before we can do that, though, we have to get our mass in grams so that its units agree with the heat capacity units:

$$\frac{11.5\ \cancel{kg}}{1} \times \frac{1{,}000\ g}{1\ \cancel{kg}} = 1.15 \times 10^4\ g$$

Now we can use Equation (2.3) :

$$q = m \cdot c \cdot \Delta T$$

$$q = (1.15 \times 10^4\ g) \cdot \left(0.837 \frac{J}{g \cdot {}^\circ C}\right) \cdot (45\ {}^\circ C - 25\ {}^\circ C)$$

$$q = (1.15 \times 10^4\ \cancel{g}) \cdot \left(0.837 \frac{J}{\cancel{g} \cdot {}^\circ \cancel{C}}\right) \cdot (2.0 \times 10^1\ {}^\circ \cancel{C}) = \underline{1.9 \times 10^5\ J}$$

12. [2 pts, one for getting ΔT and one for getting T_{final}] In order to get the copper's new temperature, we need to solve for ΔT in Equation (2.3). We can do this because we have the mass and heat given in the problem and the specific heat. Remember, though, since the copper lost heat, "q" is negative! So first we rearrange the equation to solve for ΔT:

$$\Delta T = \frac{q}{m \cdot c}$$

$$\Delta T = \frac{-856.7\ \cancel{J}}{(25.0\ \cancel{g}) \cdot \left(0.385 \frac{\cancel{J}}{\cancel{g} \cdot {}^\circ C}\right)} = -89.0\ {}^\circ C$$

Now that we have ΔT, we can rearrange Equation (2.4) to solve for final temperature:

$$T_{final} = \Delta T + T_{initial} = -89.0\ {}^\circ C + 25\ {}^\circ C = \underline{-64\ {}^\circ C}$$

13. [2 pts, one for getting q_{water} and one for getting c] The problem says to ignore the heat absorbed by the calorimeter, so $q_{calorimeter} = 0$ in Equation (2.5). We have enough information to calculate q_{water}, so we might as well start there:

$$q_{water} = m \cdot c \cdot \Delta T$$

$$q_{water} = (150.0\ \cancel{g}) \cdot \left(4.184 \frac{J}{\cancel{g} \cdot {}^\circ \cancel{C}}\right) \cdot (3.4\ {}^\circ \cancel{C}) = 2{,}100\ J$$

Now we can determine q_{metal} from Equation (2.5):

$$-q_{metal} = q_{water} + q_{calorimeter} = 2{,}100 \text{ J}$$

$$q_{metal} = -2{,}100 \text{ J}$$

We can use that value of heat and the metal's change in temperature to calculate its specific heat. However, we have to figure out the ΔT. The metal started out at 100.0 °C and ended up with the same temperature as the water. What was the final temperature of the water? Well, the water started out at 22.1 °C and its temperature rose 3.4 °C. Thus, its final temperature (which is also the metal's final temperature) is 25.5 °C.

$$c = \frac{q}{m \cdot \Delta T}$$

$$c = \frac{-2{,}100 \text{ J}}{(50.0 \text{ g}) \cdot (25.5\ ^\circ\text{C} - 100.0\ ^\circ\text{C})} = \underline{0.56 \frac{\text{J}}{\text{g} \cdot {}^\circ\text{C}}}$$

14. (2 pts, ½ point for each assumption) Dalton's atomic theory assumed four things:

1. All elements are composed of small, indivisible particles called "atoms."
2. All atoms of the same element have exactly the same properties.
3. Atoms of different elements have different properties.
4. Compounds are formed when atoms are joined together. Since atoms are indivisible, they can only join together in simple, whole-number ratios.

15. (1 pt) A compound is ionic if, when dissolved in water, it conducts electricity. If it does not conduct electricity, the compound is covalent.

16. (1 pt) If a compound has a metal in it, it must be ionic. If it has no metals, it is covalent.

17. (2 pts, one for each answer. Take one point off for every wrong compound listed.) All elements that lie to the left of the jagged line on the chart are metals, while all elements to the right of the jagged line are nonmetals. A molecule is covalent only if it has no metals in it. Thus, CO_2 and C_3H_8 are covalent.

18. (2 pts, one for determining the proper recipe and one for scaling up to 1.000×10^3 g) The reaction starts with 100.0 g + 100.0 g = 200.0 g of matter; thus, there must be 200.0 g of matter after everything is finished. According to the problem, these amounts of magnesium and bromine made 115.2 g of magnesium bromide along with left over magnesium. Since all 200.0 g must be accounted for, the remaining mass must be in the magnesium:

$$\text{Mass of magnesium} = \text{Total mass} - \text{Mass of product}$$

$$\text{Mass of magnesium} = 200.0\,\text{g} - 115.2\,\text{g} = 84.8\ \text{g}$$

By the law of mass conservation, then, there were 84.8 g of magnesium left over. Since we started out with 100.0 g of magnesium and there were 84.8 g left over, then only 100.0 g - 84.8 g = 15.2 g were actually used to make magnesium bromide. Thus, the proper recipe for making 115.2 g of magnesium bromide is to add 100.0 g of bromine to 15.2 g of magnesium.

The problem, however, asks us the recipe for making 1.000 kg, or 1.000 x 10^3 g magnesium bromide. Therefore, we need to determine how much to increase the amount of ingredients in order to make this larger amount:

$$115.2\,\text{g} \cdot \text{x} = 1.000 \times 10^3\ \text{g}$$

$$\text{x} = \frac{1.000 \times 10^3\ \text{g}}{115.2\ \text{g}} = 8.681$$

To make 1.000 kg, then, we just multiply the amount of each component by 8.681:

$$\text{Mass of bromine} = 100.0\,\text{g} \cdot 8.681 = 868.1\ \text{g}$$

$$\text{Mass of magnesium} = 15.2\,\text{g} \cdot 8.681 = 132\ \text{g}$$

You need 868.1 g of bromine and 132 g of magnesium to make 1.000 kg of magnesium bromide.

19. (1 pt each) a. P_2O_6 b. phosphorus trihydride c. NO_2 d. calcium sulfide

20. (1 pt) Since molecules move slower and are closer together in the solid phase compared to the liquid phase, the liquid must have turned into a solid.

21. (1 pt each) a. heterogeneous mixture b. compound c. element d. homogeneous mixture

22. (1 pt each) a. physical change b. chemical change c. physical change d. chemical change

23. (1 pt) $2HNO_3\ (aq) + Mg\ (s) \rightarrow MgN_2O_6\ (aq) + H_2\ (g)$

24. (2 pts) $2C_{10}H_{22} + 31O_2 \rightarrow 20CO_2 + 22H_2O$

25. (2 pts) $6CO_2\ (g) + 6H_2O\ (l) \rightarrow C_6H_{12}O_6\ (s) + 6O_2\ (g)$

Total possible points: 48

SOLUTIONS TO QUARTERLY TEST #2

1. a. (1 pt) Complete combustion, because O_2 is added while CO_2 and H_2O are produced.
   b. (1 pt) Formation, because 2 elements come together to produce 1 compound.
   c. (1 pt) None of these
   d. (1 pt) Decomposition, because a single compound is breaking down into it elements.

2. (1 pt) Complete combustion produces carbon dioxide and water. Incomplete combustion produces either carbon monoxide and water or carbon and water.

3. (2 pts, one for getting the equation and one for balancing it) The compound is made up of Na, H, S, and O. H and O are homonuclear diatomics, thus:

$$Na + H_2 + S + O_2 \rightarrow NaHSO_4$$

The Na's and S's are already balanced. To balance the H's without disturbing the others, we will use a fraction. The O's are easy to balance:

$$Na + \frac{1}{2}H_2 + S + 2O_2 \rightarrow NaHSO_4$$

Eliminating the fraction:

$$2 \text{ x } [Na + \frac{1}{2}H_2 + S + 2O_2\ ] \rightarrow 2 \text{ x } [NaHSO_4]$$

$$\underline{2Na + H_2 + 2S + 4O_2 \rightarrow 2NaHSO_4}$$

4. (2 pts, one for getting the equation and one for balancing it) Complete combustion reactions have O_2 and whatever is being burned as reactants while CO_2 and H_2O are products:

$$C_8H_{18} + O_2 \rightarrow CO_2 + H_2O$$

Balancing the C's and H's is easy:

$$C_8H_{18} + O_2 \rightarrow 8CO_2 + 9H_2O$$

There are 25 O's on the products side and 2 on the reactants side. To balance those:

$$C_8H_{18} + \frac{25}{2}O_2 \rightarrow 8CO_2 + 9H_2O$$

Eliminating the fraction:

$$\underline{2C_8H_{18} + 25O_2 \rightarrow 16CO_2 + 18H_2O}$$

5. (2 pts, one for getting the mass in amu and one for converting to kg) An $MgCO_3$ molecule has one Mg (24.3 amu each), one C (12.0 amu), and three O's (16.0 amu each), thus:

$$\text{Mass of } MgCO_3 = 1 \text{ x } 24.3 \text{ amu} + 12.0 \text{ amu} + 3 \text{ x } 16.0 \text{ amu} = 84.3 \text{ amu}$$

Since we can only relate amu to g, we need to do a two-step conversion:

$$\frac{84.3\ \cancel{\text{amu}}}{1} \times \frac{1.66 \times 10^{-24}\ \cancel{\text{g}}}{1.00\ \cancel{\text{amu}}} \times \frac{1\ \text{kg}}{1{,}000\ \cancel{\text{g}}} = \underline{1.40 \times 10^{-25}\ \text{kg}}$$

6. (2 pts, one for getting the mass in amu and one for converting to moles) The molecular mass of K_2CO_3 is 2 x 39.1 amu + 1x12.0 amu + 3x16.0 amu = 138.2 amu. This means:

$$138.2\ \text{g}\ K_2CO_3 = 1\ \text{mole}\ K_2CO_3$$

Now we do the grams to moles conversion:

$$\frac{150.0\ \cancel{\text{g}\ K_2CO_3}}{1} \times \frac{1\ \text{mole}\ K_2CO_3}{138.2\ \cancel{\text{g}\ K_2CO_3}} = \underline{1.085\ \text{moles}\ K_2CO_3}$$

7. (2 pts, one for getting the mass in amu and one for converting to grams) This is just another conversion problem, but this time we are converting moles into grams. We still need to determine the molecular mass of $MgBr_2$ first, however:

$$\text{Mass of } MgBr_2 = 1 \text{ x } 24.3 \text{ amu} + 2 \text{ x } 79.9 \text{ amu} = 184.1 \text{ amu}$$

This means:

$$184.1\ \text{grams}\ MgBr_2 = 1\ \text{mole}\ MgBr_2$$

Now we can do our conversion:

$$\frac{0.342\ \cancel{\text{moles}\ MgBr_2}}{1} \times \frac{184.1\ \text{g}\ MgBr_2}{1\ \cancel{\text{mole}\ MgBr_2}} = \underline{63.0\ \text{g}\ MgBr_2}$$

8. (1 pt) The limiting reactant is <u>KCl</u> because two moles of KCl react with one mole of $MgBr_2$. Since 2 moles of $MgBr_2$ are added, the two moles of KCl will run out and there will still be one extra mole of $MgBr_2$ left over.

9. (1pt) The chemist can only use Gay-Lussac's Law to relate the volumes of <u>HCl and CO_2</u>, since those are the only gases in the equation.

10. (2 pts, one for each answer. Take one point off for every wrong letter listed.) The formulas in <u>(b) and (c)</u> are empirical formulas, because the subscripts have no common factor. (a) is not an empirical formula because the subscripts have a common factor of 9.

11. (1 pt) Since we are using liters and the two substances in the problem are gases, we can use Gay-Lussac's Law and relate the liters of each substance:

$$3\ \text{L}\ CCl_2F_2 = 3\ \text{L}\ CCl_4$$

$$\frac{115.0\ \cancel{\text{L}\ CCl_2F_2}}{1} \times \frac{3\ \text{L}\ CCl_4}{3\ \cancel{\text{L}\ CCl_2F_2}} = \underline{115.0\ \text{L}\ CCl_4}$$

12. (3 pts, one for converting to moles of $Ca_3P_2O_8$, one for converting to moles of fertilizer, and one for converting to grams of fertilizer) This problem asks us to relate the quantity of limiting reactant to the quantity of product, but the quantities are both in grams. So first we must convert to moles:

$$\frac{150.0 \text{ g } Ca_3P_2O_8}{1} \times \frac{1 \text{ mole } Ca_3P_2O_8}{310.3 \text{ g } Ca_3P_2O_8} = 0.4834 \text{ moles } Ca_3P_2O_8$$

Now we can use stoichiometry to relate the two substances:

$$\frac{0.4834 \text{ moles } Ca_3P_2O_8}{1} \times \frac{1 \text{ mole } CaH_4P_2O_8}{1 \text{ mole } Ca_3P_2O_8} = 0.4834 \text{ moles } CaH_4P_2O_8$$

Now that we know how much fertilizer was made, we just need to get it in grams:

$$\frac{0.4834 \text{ moles } CaH_4P_2O_8}{1} \times \frac{234.1 \text{ g } CaH_4P_2O_8}{1 \text{ mole } CaH_4P_2O_8} = \underline{113.2 \text{ g } CaH_4P_2O_8}$$

13. [3 pts, one for converting to moles of Mg, one for converting to moles of water, and one for converting to grams of water] This problem asks us to relate the amount of one reactant to the amount of another. This is easy, as long as we start with moles, not grams:

$$\frac{4.1 \times 10^3 \text{ g Mg}}{1} \times \frac{1 \text{ mole Mg}}{24.3 \text{ g Mg}} = 1.7 \times 10^2 \text{ moles Mg}$$

Now we can convert from Mg to water :

$$\frac{1.7 \times 10^2 \text{ moles Mg}}{1} \times \frac{2 \text{ moles } H_2O}{1 \text{ mole Mg}} = 3.4 \times 10^2 \text{ moles } H_2O$$

Now we know how much water is needed. All we need to do now is convert to grams:

$$\frac{3.4 \times 10^2 \text{ moles } H_2O}{1} \times \frac{18.0 \text{ g } H_2O}{1 \text{ mole } H_2O} = \underline{6.1 \times 10^3 \text{ g } H_2O}$$

14. (1 pt) The mass of the empirical formula is:

Mass CH = 1 x 12.0 amu + 1 x 1.01 amu = 13.0 amu

In order to get that equal to the molecular mass, we must multiply it by 6. Thus, the empirical formula must be multiplied by 6 as well:

$$C_{1x6}H_{1x6} = \underline{C_6H_6}$$

15. (2 pts, one for getting moles of each product and one for getting the empirical formula) The unbalanced equation for the decomposition is:

$$C_xH_yO_z \rightarrow C + H_2 + O_2$$

To get the stoichiometric coefficients on the products side, we use the experimental data:

$$\frac{63.2 \text{ gC}}{1} \times \frac{1 \text{ mole C}}{12.0 \text{ gC}} = 5.27 \text{ moles C}$$

$$\frac{5.26 \text{ g}H_2}{1} \times \frac{1 \text{ mole } H_2}{2.02 \text{ g}H_2} = 2.60 \text{ moles } H_2$$

$$\frac{41.6 \text{ g}O_2}{1} \times \frac{1 \text{ mole } O_2}{32.0 \text{ g}O_2} = 1.30 \text{ moles } O_2$$

So the equation becomes:

$$C_xH_yO_z \rightarrow 5.27C + 2.60H_2 + 1.30O_2$$

We now must divide by the smallest number to make them integers:

$$C_xH_yO_z \rightarrow \frac{5.27}{1.30}C + \frac{2.60}{1.30}H_2 + \frac{1.30}{1.30}O_2$$

$$C_xH_yO_z \rightarrow 4C + 2H_2 + O_2$$

To balance the equation, then, x=4, y=4, z=2. This makes a formula of $C_4H_4O_2$, but that is not an empirical formula because the subscripts have a common factor of 2. Thus, the real empirical formula is C_2H_2O.

16. (1 pt) If a substance has an imbalance of charges, it takes on the charge which the most have. Thus, this substance will be positively charged.

17. (1 pt) Remember ROY G. BIV. This is the order of visible light wavelengths from the largest to the smallest. Thus, the red light bulb has larger wavelengths. When wavelength is large, however, frequency is small,. thus, the yellow light has the highest frequency. The higher the frequency, the higher the energy, so the yellow light also has the highest energy.

18. a. (1 pt) Looking at the chart, K has an atomic number of 19. This means it has 19 protons and 19 electrons. Its mass number, according to the problem, is 39. If it has 39 total protons + neutrons and it has 19 protons, then it has 39 - 19 = 20 neutrons.

b. (1 pt) Looking at the chart, Ne has an atomic number of 10. This means it has 10 protons and 10 electrons. Its mass number, according to the problem, is 20. If it has 20 total protons + neutrons and it has 10 protons, then it has 20 - 10 = 10 neutrons.

c. (1 pt) Looking at the chart, Mn has an atomic number of 25. This means it has 25 protons and 25 electrons. Its mass number, according to the problem, is 52. If it has 52 total protons + neutrons and it has 25 protons, then it has 52 - 25 = 27 neutrons.

d. (1 pt) Looking at the chart, I has an atomic number of 53. This means it has 53 protons and 53 electrons. Its mass number, according to the problem, is 131. If it has 131 total protons + neutrons and it has 53 protons, then it has 131 - 53 = 78 neutrons.

19. a. (1 pt) To get to element Fe, we must go through row 1, which has two boxes in the s orbital block ($1s^2$). We then go through all of row 2 which has two boxes in the s orbital block and six boxes in the p orbital block ($2s^22p^6$). We also go through row 3, which has two boxes in the s orbital block and six in the p orbital block ($3s^23p^6$). We then go to the fourth row where we pass through both boxes in the s orbital block ($4s^2$). Finally, we go through six boxes in the d orbital block. Since we subtract one from the row number for d orbitals, this gives us $3d^6$. Thus, our final electron configuration is:

$$1s^22s^22p^63s^23p^64s^23d^6$$

b. (1 pt) To get to element P, we must go through row 1, which has two boxes in the s orbital block ($1s^2$). We then go through all of row 2 which has two boxes in the s orbital block and six boxes in the p orbital block ($2s^22p^6$). We also go through both boxes in the s orbital block of row 3, ($3s^2$). Finally, we go through three boxes in the p orbital block of row 3, giving us $3p^3$. Thus, our final electron configuration is:

$$1s^22s^22p^63s^23p^3$$

c. (1 pt) To get to element K, we must go through row 1, which has two boxes in the s orbital block ($1s^2$). We then go through all of row 2 which has two boxes in the s orbital block and six boxes in the p orbital block ($2s^22p^6$). We also go through row 3, which has two boxes in the s orbital block and six in the p orbital block ($3s^23p^6$). We then go to the fourth row where we go through the first box, giving us $4s^1$. Thus, our final electron configuration is:

$$1s^22s^22p^63s^23p^64s^1$$

20. a. (1 pt) The nearest 8A element that has a lower atomic number than Cl is Ne. The only difference between Cl and Ne is that there are two boxes in the row 3, s orbital group and five boxes in the row 3, p orbital group, and thus, the abbreviated electron configuration for Cl is:

$$[Ne]3s^23p^5$$

b. (1 pt) The nearest 8A element that has a lower atomic number than Nb is Kr. The only difference between Nb and Kr is that there are two boxes in the row 5, s orbital group and three boxes in the row 5, d orbital group. Thus, the abbreviated electron configuration for Nb is:

$$[Kr]5s^24d^3$$

c. (1 pt) The nearest 8A element that has a lower atomic number than Cs is Xe. The only difference between Cs and Xe is that there is one box in the row 6, s orbital group. Thus, the abbreviated electron configuration for Cs is:

$$[Xe]6s^1$$

21. (1 pt) Metals tend to give up electrons to attain the ideal electron configuration, while nonmetals tend to gain electrons for the same purpose.

22. (1 pt) a. Be is in group 2A, so it has 2 valence electrons:

Be•
•

b. (1 pt) C is in group 4A, so it has 4 valence electrons:

•
•C•
•

c. (1 pt) Cl is in group 7A, so it has 7 valence electrons:

•
:Cl:
••

23. a. (1 pt) Al is in group 3A, so it wants a charge of 3+. Sulfur is in group 6A, so sulfide will have a charge of 2-. Ignoring the signs and switching the numbers gives us $\underline{Al_2S_3}$.

b. (1 pt) K is in group 1A, so it wants a charge of 1+. Nitrogen is in group 5A, so nitride will have a charge of 3-. Ignoring the signs and switching the numbers gives us $\underline{K_3N}$.

c. (1 pt) Mg is in group 2A, so it wants a charge of 2+. Sulfur is in group 6A, so sulfide will have a charge of 2-. The numbers are the same so we ignore them: $\underline{MgS}$.

d. (1 pt) Cr is an exception, because there is a Roman numeral in the name. The numeral tells us that Cr want a charge of 3+. Oxygen is in group 6A, so chloride will have a charge of 2-. Ignoring the signs and switching the numbers gives us $\underline{Cr_2O_3}$.

24. (2 pts) The chemical formula tells us that we have one N and three H's to work with:

•
•N: H• H• H•
•

Because N has the most unpaired electrons, it goes in the center and we try to attach the H's to it. This is easy since each H has a space for an unpaired electron, and the N has 3 unpaired electrons. The Lewis structure, then, looks like this:

H
••
H:N:
••
H

All atoms have their ideal electron configuration, so we are all set. Now we just have to replace the shared electron pairs with dashes:

H
|
H–N:
|
H

25. (2 pts) The chemical formula tells us that we have two H's, one C, and one O to work with:

H• H• •Ċ• •Ȯ:

The C has the most unpaired electrons, so it goes in the middle. We attach the others to it:

H:C:O:
H

The H's now have two electrons, so they are all set. The C and O, however, have only seven each. We will give the O eight by taking the unpaired electron on the C and putting it between the C and the O. We will also give the C its eight by taking the unpaired electron on the oxygen and moving it between the C and the O:

H:C::O:
H

Now all atoms have 8 valence electrons. All we have to do is replace the shared electron pairs with dashes:

H–C=O:
|
H

Total possible points: 53

SOLUTIONS TO QUARTERLY TEST #3

1. Ionic compounds are named by simply listing the ions present. In order to get the formula, you must determine the charge of each ion and balance those charges.

a. (1 pt) The name indicates a potassium ion and a carbonate ion. Potassium is abbreviated with K, and, since it is in group 1A, it has a charge of 1+. We are supposed to have memorized that the carbonate ion is CO_3^{2-}. Ignoring the signs and switching the numbers gives us:

K_2CO_3

b. (1 pt) The name indicates a calcium ion and a nitrate ion. Calcium is abbreviated with Ca, and, since it is in group 2A, it has a charge of 2+. We are supposed to have memorized that the nitrate ion is NO_3^-. Ignoring the signs and switching the numbers gives us:

$Ca(NO_3)_2$

c. (1 pt) The name indicates an aluminum ion and a sulfate ion. Aluminum is abbreviated with Al, and, since it is in group 3A, it has a charge of 3+. We are supposed to have memorized that the sulfate ion is SO_4^{2-}. Ignoring the signs and switching the numbers gives us:

$Al_2(SO_4)_3$

d. (1 pt) The name indicates a calcium ion and a chromate ion. Calcium is abbreviated with Ca, and, since it is in group 2A, it has a charge of 2+. We are supposed to have memorized that the chromate ion is CrO_4^{2-}. The charges are the same so we ignore them:

$CaCrO_4$

2. In order to name ionic compounds, we only have to put the names of the ions together.

a. (1 pt) Since we see that PO_4 is in parentheses, that means it is a polyatomic ion. We are supposed to have memorized that PO_4^{3-} is the phosphate ion, and the only other ion is the single-atom calcium ion. Thus, the name is calcium phosphate.

b. (1 pt) In looking at this molecule, we should notice the NO_3. It tells us the nitrate polyatomic ion is present. The only thing left after that is the calcium ion. Thus, the name is calcium nitrate.

c. (1 pt) We should recognize NH_4 as a polyatomic ion. We are supposed to have memorized that NH_4^+ is the ammonium ion, and the only other ion is the single-atom chloride ion. Thus, the name is ammonium chloride.

d. (1 pt) In looking at this molecule, we should notice the CO_3. It tells us that the carbonate polyatomic ion is present. The only thing left after that is the potassium ion. Thus, the name is potassium carbonate.

3. (1 pt) To determine shape, we must first draw the Lewis structure:

:C≡O:

Since there are only two atoms here, the molecule is linear with a bond angle of 180°. The picture looks just like the Lewis structure.

4. (2 pts, one for the shape and bond angle and one for the picture) To determine shapes, we must first draw the Lewis structure:

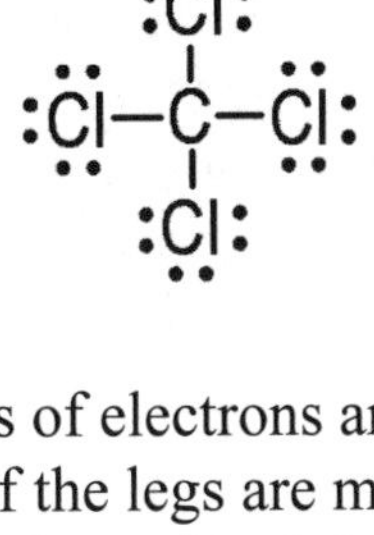

We see that the central atom has four groups of electrons around it. Since there are four groups, the basic shape is that of a tetrahedron. None of the legs are missing because the molecule contains no non-bonding pairs of electrons. As a result, the molecule's shape is tetrahedral with a bond angle of 109°:

5. (2 pts, one for the shape and bond angle and one for the picture) To determine shapes, we must first draw the Lewis structure:

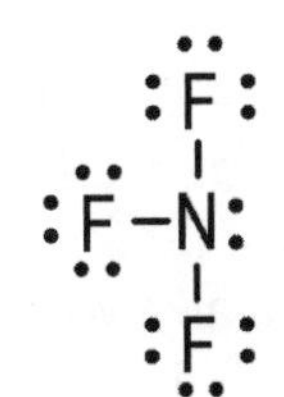

We see that the central atom has four groups of electrons around it. Three of them are bonds, and one is a non-bonding pair. Since there are four groups, the basic shape is that of a tetrahedron. However, one of the legs is missing because it contains a nonbonding pair of electrons. As a result, the molecule's shape is pyramidal with a bond angle of 107°:

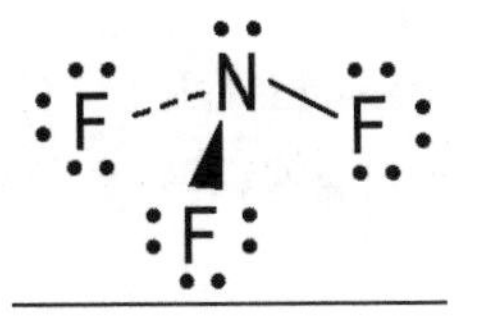

6. (1 pt) To determine shapes, we must first draw the Lewis structure:

$$:\underset{..}{S}=C=\underset{..}{S}:$$

We see that the central atom has 2 groups of electrons around it. Since there are 2 groups, the shape is linear with a bond angle of 180^0. The picture looks just like the Lewis structure.

7. (2 pts, one for each answer. Take a point away for every wrong formula listed.) Purely covalent molecules are those with no polar bonds or those whose geometric symmetry allows the bonds to cancel each other out. CCl_4 is purely covalent because the polar bonds are all pointed in precisely opposite directions. Thus, the electron pull of one bond is canceled out by the others. In the same way, CS_2 is purely covalent, because the polar bonds are pointed in precisely the opposite directions.

8. (1 pt) Acids donate H^+ ions. In this reaction, C_2H_6O becomes $C_2H_5O^-$. The only way that can happen is if it gives up an H^+. Thus, C_2H_6O is the acid.

9. (1 pt) Looking at the chemical formula, you should immediately see the sulfate ion (SO_4^{2-}). The positive ion is the aluminum ion. Aluminum's position on the chart tells you it's a 3+ ion. Thus, the molecule splits into two Al^{3+} ions and three SO_4^{2-} ions.

10. (2 pts, one for the unbalanced equation and one for the balanced equation) Magnesium hydroxide is $Mg(OH)_2$. Acids and bases usually react to give a salt and water. The salt is made up of the positive ion from the base (Mg^{2+}) and the negative ion left over when the acid gets rid of its H^+ ions. In this case, that will be NO_3^-. These two ions form $Mg(NO_3)_2$. The unbalanced equation, then, is:

$$HNO_3 + Mg(OH)_2 \rightarrow H_2O + Mg(NO_3)_2$$

Now all we have to do is balance it:

$$2HNO_3 + Mg(OH)_2 \rightarrow 2H_2O + Mg(NO_3)_2$$

11. (1 pt) In this case, the base does not contain an hydroxide ion. Thus, a salt and water are not formed in this problem. Here, we just rely on the definition of acids and bases. PH_3 will want to gain an H^+ to become PH_4^+, and the nitric acid will want to give up its H^+ ion to become NO_3^-.

$$HNO_3 + PH_3 \rightarrow PH_4^+ + NO_3^-$$

12. (2 pts, one for the concentration of the stock KOH solution and one for the dilution instructions) The last part of this problem is a dilution problem, but in order to get the original concentration, we must deal with the first part of the problem, which is a concentration problem.

$$\frac{1050.0 \text{ \sout{g KOH}}}{1} \times \frac{1 \text{ mole KOH}}{56.1 \text{ \sout{g KOH}}} = 18.7 \text{ moles KOH}$$

$$\text{Concentration} = \frac{\#\text{moles}}{\#\text{liters}} = \frac{18.7 \text{ moles KOH}}{1.50 \text{ liters}} = 12.5 \text{ M}$$

Now that we know the concentration of the stock solution, this is just a dilution problem:

$$M_1V_1 = M_2V_2$$

$$12.5\ M \cdot V_1 = 0.10\ M \cdot 200.0\ mL$$

$$V_1 = \frac{0.10\,\cancel{M} \cdot 200.0\,mL}{12.5\,\cancel{M}} = 1.6\ mL$$

The chemist needs to take 1.6 mL of the stock solution and mix it with enough water to make 200.0 mL of solution.

13. (3 pts, one for getting moles of HCl, one for converting to moles of KOH, and one for getting the concentration of KOH) Remember, titrations are just stoichiometry problems, so first we have to come up with a balanced chemical equation:

$$HCl + KOH \rightarrow KCl + H_2O$$

Since the endpoint was reached, we know that there was exactly enough acid added to eat up all of the base. First, then, we calculate how many moles of acid were added:

$$\frac{2.5\ moles\ HCl}{1\ \cancel{L}} \times \frac{0.0147\ \cancel{L}}{1} = 0.037\ moles\ HCl$$

We can now use the chemical equation to determine how many moles of base were present:

$$\frac{0.037\ \cancel{moles\ HCl}}{1} \times \frac{1\ mole\ KOH}{1\ \cancel{mole\ HCl}} = 0.037\ moles\ KOH$$

Now that we have the number of moles of base present, we simply divide by the volume of acid to get concentration:

$$Concentration = \frac{\#\ moles}{\#\ liters} = \frac{0.037\ moles\ NaOH}{0.0500\ liters} = \underline{0.74\ M}$$

14. (1 pt) The solubility of gases increase by increased pressure and decreased temperature.

15. (1 pt) The freezing point depression depends on the solvent (which is water in each case), the molality (which is the same in each case), and the number of particles that the solute splits up into when it dissolves. In this problem, $Al(NO_3)_3$ splits into the most ions (4), so it will cause the greatest freezing point depression.

16. (3 pts, one for getting the moles of aluminum nitrate, one for getting the moles of aluminum sulfate, and one for getting the grams of aluminum sulfate) This is just a stoichiometry problem. We can tell this by the fact that we are being asked to determine the amount of one substance when we are

given the amount of another substance. The only way to do that is by stoichiometry. Now, in order to do stoichiometry, we must first get our amount in moles.

$$\frac{1.25 \text{ moles } Al(NO_3)_3}{1 \not{L}} \times 0.191 \not{L} = 0.239 \text{ moles } Al(NO_3)_3$$

Now that we have moles, we can do stoichiometry:

$$\frac{0.239 \not{\text{moles } Al(NO_3)_3}}{1} \times \frac{1 \text{ mole } Al_2(SO_4)_3}{2 \not{\text{mole } Al(NO_3)_3}} = 0.120 \text{ moles } Al_2(SO_4)_3$$

Now, of course, this is not quite the answer we need. We were asked to figure out how many grams of aluminum sulfate were produced, so we have to convert from moles back to grams:

$$\frac{0.120 \not{\text{moles } Al_2(SO_4)_3}}{1} \times \frac{342.3 \text{ grams } Al_2(SO_4)_3}{1 \not{\text{mole } Al_2(SO_4)_3}} = \underline{41.1 \text{ grams } Al_2(SO_4)_3}$$

17. (1 pt) Freezing point depression is determined by Equation (11.2). We are already given 2 of the 4 variables in the equation (K_f,ΔT), and we can calculate a third (i). Molality is the only unknown, so we can solve for it. Na_2CO_3, since it is ionic, will split up into ions (two sodium ions and one carbonate ion), so i= 3:

$$\Delta T = -i \cdot K_f \cdot m$$

$$-11\ ^{\circ}C = -3 \cdot 1.86 \frac{^{\circ}C}{\text{molal}} \cdot m$$

$$m = \frac{-11\ ^{\circ}\not{C}}{-3 \cdot 1.86 \frac{^{\circ}\not{C}}{\text{molal}}} = \underline{2.0 \text{ molal}}$$

18. (1 pt) Molality is defined as the number of moles of solute per kg of solvent. Thus, 2.0 m is 2.0 moles of Na_2CO_3 per 1.00 kg of solvent. We don't have 1.00 kg, we have 0.5000 kg, so we first need to calculate how many moles are required:

$$\frac{2.0 \text{ moles } Na_2CO_3}{1.0 \not{\text{kg water}}} \times 0.5000 \not{\text{kg water}} = 1.0 \text{ mole } Na_2CO_3$$

That's how many moles are required, but we need to know grams. That's an easy conversion:

$$\frac{1.0 \not{\text{moles } Na_2CO_3}}{1} \times \frac{106 \text{ g } Na_2CO_3}{1 \not{\text{mole } Na_2CO_3}} = \underline{110 \text{ g } Na_2CO_3}$$

19. (2 pts, one for the molality and one for the boiling point) To calculate boiling points, we must use Equation (11.3). To do that, however, we must know "i" and "m". To calculate "m":

$$\frac{100.0\ \cancel{\text{g NaCl}}}{1} \times \frac{1\ \text{mole NaCl}}{58.5\ \cancel{\text{g NaCl}}} = 1.71\ \text{moles NaCl}$$

$$\text{m} = \frac{1.71\ \text{moles NaCl}}{1.1000\ \text{kg water}} = 1.55\ \text{m}$$

Since sodium chloride is an ionic compound, it dissolves by splitting up into two ions; thus, i = 2.

$$\Delta T = i \cdot K_b \cdot m = 2 \cdot 0.512 \frac{^{\circ}\text{C}}{\cancel{\text{m}}} \cdot 1.55\cancel{\text{m}} = 1.59\ ^{\circ}\text{C}$$

This means that the boiling point of the solution is 1.59 °C *higher* than that of pure water. The boiling point of pure water is 100.0 °C, so the boiling point of this solution is 101.6 °C.

20. (1 pt, one-third of a point for each property) The molecules (or atoms) that make up an ideal gas must be small compared to the volume available to the gas. The gas molecules (or atoms) must be so far apart that they do not attract or repel each other. Also, all collisions that occur must be elastic.

21. (1 pt, ½ point for the pressure and ½ point for the temperature) Gases behave ideally when their pressure is near or lower than 1.00 atm and when their temperature is near or higher than 273K.

22. (2 pts, one if all of the units match and temperature is in K, and one for the new volume) This is obviously a combined gas law problem, with P_1 = 790 torr, V_1 = 156 mL, T_1 = 298 K, P_2 = 1.00 atm (standard pressure), and T_2 = 273 K (standard temperature). The problem asks us to determine the new volume, so we have to rearrange Equation (12.10) to solve for V_2:

$$\frac{P_1 V_1 T_2}{T_1 P_2} = V_2$$

Before we can plug in the numbers, however, we need to convert T_1 to Kelvin. Additionally, we need to make the pressure units the same. We can do this by converting torr into atm or vice versa. I will choose to do the latter:

$$P_2 = \frac{1.00\ \text{atm}}{1} \times \frac{760\ \text{torr}}{1\ \text{atm}} = 7.60 \times 10^2\ \text{torr}$$

Now we can plug in the numbers:

$$\frac{790\ \cancel{\text{torr}} \cdot 156\ \text{mL} \cdot 273\cancel{\text{K}}}{298\cancel{\text{K}} \cdot 7.60 \times 10^2\ \cancel{\text{torr}}} = V_2$$

$$V_2 = 150\ \text{mL}$$

23. (2 pts, one for getting the mole fractions and one for getting the partial pressures) Remember, the partial pressure of the individual gases in a mixture depends on the mole fraction of each gas, according to Dalton's Law. Mole fraction is defined as the number of *moles* of component divided by the total number of moles. Right now, the problem gives us *grams*, not moles. Thus, we must first convert from grams to moles:

$$\frac{1.2\ \cancel{g N_2}}{1} \times \frac{1\text{ mole } N_2}{28.0\ \cancel{g N_2}} = 0.043\text{ moles } N_2$$

$$\frac{1.5\ \cancel{g O_2}}{1} \times \frac{1\text{ mole } O_2}{32.0\ \cancel{g O_2}} = 0.047\text{ moles } O_2$$

Now that we have the number of moles of each component, we can calculate the total number of moles in the mixture:

$$\text{Total number of moles} = 0.043\text{ moles} + 0.047\text{ moles} = 0.090\text{ moles}$$

Plugging that into Equation (12.12):

$$X_{N_2} = \frac{0.043\ \cancel{\text{moles}}}{0.090\ \cancel{\text{moles}}} = 0.48$$

$$X_{O_2} = \frac{0.047\ \cancel{\text{moles}}}{0.090\ \cancel{\text{moles}}} = 0.52$$

These are not the answers yet! The question asks for partial pressure. Well, according to Dalton's Law:

$$P_1 = X_1 \cdot P_T$$

$$P_{N_2} = 0.48 \cdot 1.0\text{ atm} = \underline{0.48\text{ atm}}$$

$$P_{O_2} = 0.52 \cdot 1.0\text{ atm} = \underline{0.52\text{ atm}}$$

24. (1 pt) In this problem, we are given pressure and temperature and the mass (from which we can get moles). We are then asked to calculate V. We can do this by rearranging the ideal gas law:

$$PV = nRT$$

$$V = \frac{nRT}{P}$$

In order for this to work, though, our temperature must be in Kelvin. Also, we do not have n yet. We do, however, have mass, so we can convert it into moles:

$$\frac{15.0\ \cancel{g\ CO_2}}{1} \times \frac{1\text{ mole } CO_2}{44.0\ \cancel{g\ CO_2}} = 0.341\text{ moles } CO_2$$

Now that we have all of the correct units, we can plug the numbers into the equation.

$$V = \frac{0.341\ \cancel{\text{moles}} \cdot 0.0821 \frac{L \cdot \cancel{atm}}{\cancel{\text{mole}} \cdot \cancel{K}} \cdot 273\ \cancel{K}}{1.00\ \cancel{atm}} = \underline{7.64\text{ L}}$$

25. (3 pts, one for getting the moles of H_2O_2, one for getting the moles of water, and one for getting the volume of water). In this stoichiometry problem, we are given the amount of limiting reactant and asked to calculate how much product will be made. We start by converting the amount of limiting reactant to moles:

$$\frac{110.0\ \cancel{g\ H_2O_2}}{1} \times \frac{1\text{ mole } H_2O_2}{34.0\ \cancel{g\ H_2O_2}} = 3.24\text{ moles } H_2O_2$$

We can then use stoichiometry to determine the number of moles of H_2O produced:

$$\frac{3.24\ \cancel{\text{moles } H_2O_2}}{1} \times \frac{8\text{ moles } H_2O}{7\ \cancel{\text{moles } H_2O_2}} = 3.70\text{ moles } H_2O$$

Now we need to use the ideal gas law:

$$V = \frac{3.70\ \cancel{\text{moles}} \cdot 0.0821 \frac{L \cdot \cancel{atm}}{\cancel{\text{mole}} \cdot \cancel{K}} \cdot 614\ \cancel{K}}{2.1\ \cancel{atm}} = \underline{89\text{ L}}$$

Total possible points: 45

SOLUTIONS TO QUARTERLY TEST #4

1. (1 pt) Elements in their natural form have ΔH_f^o of zero. The only substance that is an element in its natural form is N_2 (g).

2. (3 pts, one for the position of the reactants, one for the height of the hump, and one for the position of the products) If the reactants have an energy of 20.0 kJ, the graph must start at 20.0 kJ. If the ΔH is 20.0 kJ, then the graph must end 20.0 kJ higher than it started, so it must end at 40.0 kJ. If the activation energy is 35.0 kJ, then the hump in the middle must be 35.0 kJ higher than the reactants, so it must be at an energy of 55.0 kJ:

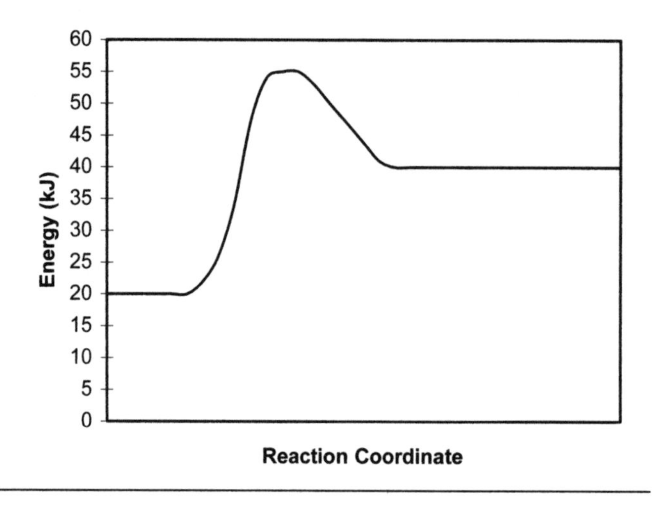

3. (1 pt) The only difference would be that the hump in the reaction with the catalyst would be smaller. That's what a catalyst does to the energy diagram. It lowers the activation energy and thus lowers the height of the hump.

4. (1 pt) Yes. For a reaction to be spontaneous, its ΔG must be negative. If the chemicals decrease in disorder, this just means ΔS is negative. A negative ΔS tends to make ΔG positive, but if ΔH is negative enough, ΔG can still be negative.

5. (1 pt) Notice that there is a gas on the products side of the equation and no gases on the reactants side. Since the gas phase is by far the most disordered, this reaction results in an *increase* in the disorder of the chemicals. Thus, ΔS is positive.

6. (2 pts, one for setting up the equation with consistent units and one for solving it properly) This problem gives us ΔH and ΔS and asks us at what temperature the reaction is spontaneous. In other words, we need to see for what temperatures ΔG is negative. Thus:

$$\Delta H - T\Delta S < 0$$

In order to use this equation, though, we need to get our units consistent:

$$\frac{194 \cancel{J}}{\text{mole} \cdot \text{K}} \times \frac{1 \text{ kJ}}{1{,}000 \cancel{J}} = 0.194 \frac{\text{kJ}}{\text{mole} \cdot \text{K}}$$

Now we can use the equation:

$$423 \frac{\text{kJ}}{\text{mole}} - \text{T}\cdot(0.194\frac{\text{kJ}}{\text{mole}\cdot\text{K}}) < 0$$

$$-\text{T}\cdot(0.194\frac{\text{kJ}}{\text{mole}\cdot\text{K}}) < -423 \frac{\text{kJ}}{\text{mole}}$$

$$\text{T}\cdot(0.194\frac{\text{kJ}}{\text{mole}\cdot\text{K}}) > 423 \frac{\text{kJ}}{\text{mole}}$$

$$\text{T} > \frac{423 \frac{\cancel{\text{kJ}}}{\cancel{\text{mole}}}}{0.194\frac{\cancel{\text{kJ}}}{\cancel{\text{mole}}\cdot\text{K}}}$$

$$\underline{\text{T} > 2{,}180 \text{ K}}$$

Notice that I had to switch the direction of the inequality sign when I multiplied both sides by a negative number. That's a rule when doing algebra on equations with inequality signs.

7. (1 pt) The rate of a chemical reaction must be M/s. Since the rate constant multiplies the concentrations of the reactants raised to their orders, the overall order can be determined by figuring out what unit must be multiplied by $1/M^3s$ to get M/s. To do this, we would have to multiply by M^4. Thus concentration must be raised to the fourth power in the rate equation. This means that the overall order must be 4.

8. (1 pt) If the order of the reaction with respect to a reactant is zero, then the reaction rate does not depend on the concentration of that reactant. Thus, nothing would happen to the rate.

9. (3 pts, one for each order, and one for k) The rate equation for this reaction will look like:

$$R = k[CO]^x[O_2]^y$$

To figure out k, x, and y, we have to look at the data from the experiment. The value for x can be determined by comparing two trials in which the concentration of CO changes, but the concentration of O_2 stays the same. This would correspond to trials 1 and 2. In these two trials, the concentration of CO doubled, and the rate doubled. This means that x = 1. The value for y can be determined by looking at trials 2 and 3, where the concentration of CO stayed the same but the concentration of O_2 doubled. When that happened, the rate increased by a factor of 4. This means y = 2, because the only way you can get a 4-fold increase in rate from a doubling of the concentration is by squaring the concentration. Thus, the rate equation becomes:

$$R = k[CO][O_2]^2$$

Now that we have x and y, we only need to find out the value for k. We can do this by using any one of the trials in the experiment and plugging the data into the equation. The only unknown will be k, and we can therefore solve for it:

$$R = k[CO][O_2]^2$$

$$0.113\,\frac{M}{s} = k \cdot (0.150\ M) \cdot (0.150\ M)^2$$

$$k = \frac{0.113\,\frac{\cancel{M}}{s}}{(0.150\ \cancel{M}) \cdot (0.0225\ M^2)} = 33.5\,\frac{1}{M^2 \cdot s}$$

Thus, the final rate equation is:

$$\underline{R = (33.5\,\frac{1}{M^2 \cdot s}) \cdot [CO][O_2]^2}$$

10. (1 pt) Since chemical reaction rate doubles for every 10 °C increment, it falls by a factor of 2 if you lower the temperature by 10 °C. If the temperature is lowered another 10 °C, it would drop by another factor of 2, making the total decrease factor of 4. Thus, to decrease the reaction rate by a factor of 4, I just lower the temperature by 20 degrees. Therefore, the new temperature <u>5 °C</u>.

11. (1 pt) <u>I^-</u> is used up in the first reaction and re-made in the second reaction. Thus, it is the catalyst.

12. (1 pt) It is a <u>homogeneous catalyst</u>, because it has the same phase as the reactants.

13. (1 pt) <u>A pH of 7 is considered neutral</u>.

14. (2 pts, subtract 1 if A is listed) Acidic solutions have pH under 7. Thus, <u>B and C are acidic</u>.

15. (1 pt) A small acid ionization constant indicates a weak acid. The higher the pH (below 7), the weaker the acid. Thus, <u>solution B</u> has the acid with the smallest ionization constant.

16. (2 pts, count off 1 if the student did not ignore the solids) To get the equation for the equilibrium constant, we have to realize that we ignore both $Pb(NO_3)_2$ and PbO, because they are solids. Thus, the equilibrium constant is:

$$K = [NO_2]_{eq}^4[O_2]_{eq}$$

Since we ignore solids, we ignore the amounts of both given in the problem. We only use the equilibrium concentrations of the gases. Plugging those equilibrium concentrations into the equation:

$$K = (0.25\ M)^4(0.019\ M) = \underline{7.4 \times 10^{-5}\ M^5}$$

17. (2 pts, one for doing the calculation, and one for predicting the shift) The equation for the equilibrium constant here is

$$K = \frac{[SO_2]_{eq}^2[O_2]_{eq}}{[SO_3]_{eq}^2}$$

If the concentrations are, in fact, equilibrium concentrations, then the equation should equal the value given for K.

$$K = \frac{(0.480\ \cancel{M})^2 \cdot (0.561\ M)}{(0.220\ \cancel{M})^2} = 2.67\ M$$

This is larger than the equilibrium constant. To get the results of the equation to equal the equilibrium point, the reaction will have to get rid of products and increase the concentration of reactants. Thus, <u>the reaction is NOT at equilibrium and will shift towards the reactants</u>.

18 a. (1 pt) The reaction is exothermic, because ΔH is negative. This tells us that energy is a product. Raising the temperature, then, will be like increasing the amount of product. This will shift the equilibrium to the reactants, consuming products. Thus, <u>the concentration of NH_4^+ will decrease</u>.

b. (1 pt) Decreasing the NH_3 concentration will shift the reaction towards the reactants. Thus, <u>the concentration of NH_4^+ will decrease</u>.

19. (1 pt) The ionization constant is simply the equilibrium constant for the acid ionization reaction. In order to determine the ionization reaction, you simply take the acid in its aqueous phase and remove an H^+. When you remove an H^+ from HI, you are left with I^-. In the end, then, the aqueous acid is the reactant, and the H^+ and I^- (both in aqueous phase) will be the products:

$$HI\ (aq) \leftrightharpoons H^+\ (aq) + I^-\ (aq)$$

The equilibrium constant for this reaction is the ionization constant, K_a:

$$K_a = \frac{[H^+][I^-]}{[HI]}$$

20. (1 pt) Na is in group 1A, so its oxidation number is always +1. We have no rule for Mn. Oxygen is almost always -2, so that's a safe bet here. That means Mn must be +7 in order for the oxidation numbers to add to zero. Thus, the oxidation numbers are: <u>Na: +1, Mn: +7, O: -2</u>.

21. (1 pt) To go from -2 to -5, you must gain negatives. This means the atom gained electrons, indicating that <u>it was reduced</u>. To go from -2 to -5, you must gain <u>3 electrons</u>.

22. (1 pt) To go from 0 to +2, you must lose negatives. This means the atom lost electrons, indicating that <u>it was oxidized</u>. To go from 0 to +2, you must lose <u>2 electrons</u>.

23. (1 pt) The only reaction in which atoms changed oxidation numbers is <u>reaction b</u>. V went from +5 to +4 and Zn went from 0 to +2.

24. (1 pt) Since V went from +5 to +4, it gained a negative. This means it gained an electron and thus V was reduced. Since Zn went from 0 to +2, it lost negatives. This means it lost electrons and thus Zn was oxidized.

25. (2 pts) In this reaction, Cu^{2+} is going from an oxidation number of +2 to an oxidation number of 0. This indicates that it is gaining electrons. Thus, the solution holding aqueous Cu^{2+} will have electrons flowing into it. Electrons flow towards the Cu^{2+}, so that container is positive (it attracts electrons) and thus will be the cathode. The Co is going from an oxidation number of 0 to an oxidation number of +2. This means it loses electrons. Since it is losing electrons, the electrons are flowing away from the container holding the solid Co. This makes that container the negative side of the battery (it repels electrons), and it is thus the anode. The picture, then, looks like this:

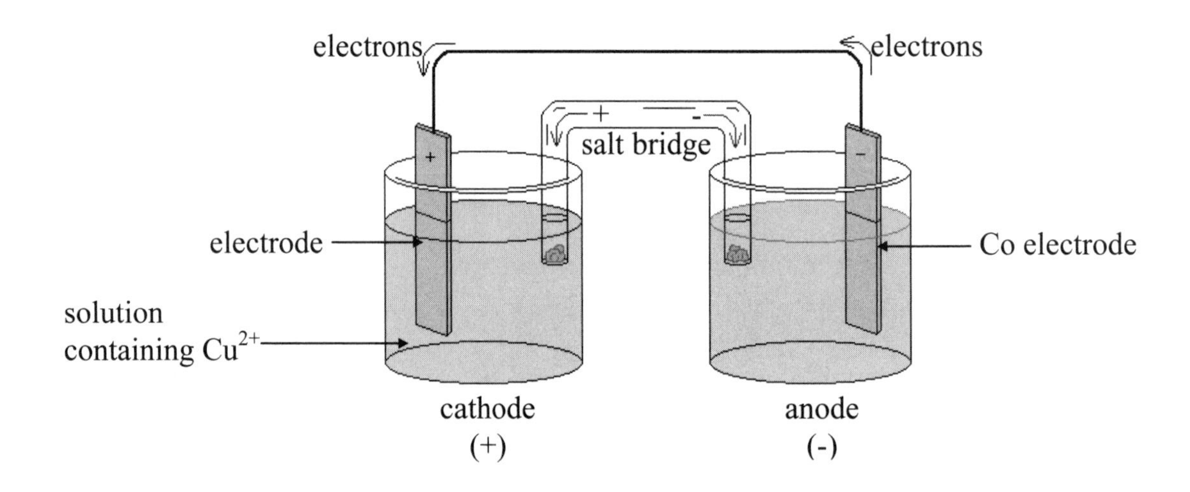

Total possible points: 35